Matrix Inequalities
and Their Extensions
to Lie Groups

Matrix Inequalities and Their Extensions to Lie Groups

Tin-Yau Tam
Xuhua Liu

CRC Press
Taylor & Francis Group
Boca Raton London New York

CRC Press is an imprint of the
Taylor & Francis Group, an **informa** business

CRC Press
Taylor & Francis Group
6000 Broken Sound Parkway NW, Suite 300
Boca Raton, FL 33487-2742

© 2018 by Taylor & Francis Group, LLC
CRC Press is an imprint of Taylor & Francis Group, an Informa business

No claim to original U.S. Government works

International Standard Book Number-13: 978-1-4987-9616-3 (Hardback)

This book contains information obtained from authentic and highly regarded sources. Reasonable efforts have been made to publish reliable data and information, but the author and publisher cannot assume responsibility for the validity of all materials or the consequences of their use. The authors and publishers have attempted to trace the copyright holders of all material reproduced in this publication and apologize to copyright holders if permission to publish in this form has not been obtained. If any copyright material has not been acknowledged please write and let us know so we may rectify in any future reprint.

Except as permitted under U.S. Copyright Law, no part of this book may be reprinted, reproduced, transmitted, or utilized in any form by any electronic, mechanical, or other means, now known or hereafter invented, including photocopying, microfilming, and recording, or in any information storage or retrieval system, without written permission from the publishers.

For permission to photocopy or use material electronically from this work, please access www.copyright.com (http://www.copyright.com/) or contact the Copyright Clearance Center, Inc. (CCC), 222 Rosewood Drive, Danvers, MA 01923, 978-750-8400. CCC is a not-for-profit organization that provides licenses and registration for a variety of users. For organizations that have been granted a photocopy license by the CCC, a separate system of payment has been arranged.

Trademark Notice: Product or corporate names may be trademarks or registered trademarks, and are used only for identification and explanation without intent to infringe.

Visit the Taylor & Francis Web site at
http://www.taylorandfrancis.com

and the CRC Press Web site at
http://www.crcpress.com

Printed and bound in Great Britain by
TJ International Ltd, Padstow, Cornwall

To Yuet-Ngan Liu (廖月顏), Tin-Yau Tam's mother, Xuedun Liu (劉學敦), Xuhua Liu's father, and our wives Kitty Tam and Joy Liu

Contents

Preface		ix
1 Review of Matrix Theory		**1**
1.1	Matrix Decompositions	3
	1.1.1 Polar Decompositions	3
	1.1.2 Singular Value Decomposition	4
	1.1.3 QR Decomposition	4
	1.1.4 Cholesky Decomposition	5
	1.1.5 Additive Decompositions	5
	1.1.6 Jordan Decompositions	6
	1.1.7 LU Decomposition	9
	1.1.8 $L\omega U$ Decomposition	10
1.2	Majorizations	13
1.3	Matrix Norms	19
1.4	The Matrix Exponential Map	23
1.5	Compound Matrices and Applications	25
	1.5.1 Compound Matrices	26
	1.5.2 Additive Compound Matrices	27
	1.5.3 Applications to Matrix Inequalities	29
2 Structure Theory of Semisimple Lie Groups		**41**
2.1	Smooth Manifolds	41
2.2	Lie Groups and Their Lie Algebras	44
2.3	Complex Semisimple Lie Algebras	48
2.4	Real Forms	49
2.5	Cartan Decompositions	51
2.6	Root Space Decomposition	55
2.7	Iwasawa Decompositions	57
2.8	Weyl Groups	59
2.9	KA_+K Decomposition	60
2.10	Complete Multiplicative Jordan Decomposition	61
2.11	Kostant's Preorder	65

3 Inequalities for Matrix Exponentials — 67

- 3.1 Golden-Thompson Inequality — 67
- 3.2 Araki-Lieb-Thirring Inequality — 75
- 3.3 Bernstein Inequality — 76
- 3.4 Extensions to Lie Groups — 80

4 Inequalities for Spectral Norm — 91

- 4.1 Matrix Inequalities for Spectral Norm — 91
- 4.2 Extensions to Lie Groups — 96

5 Inequalities for Unitarily Invariant Norms — 103

- 5.1 Matrix Inequalities for Unitarily Invariant Norms — 103
- 5.2 Extensions to Lie Groups — 105

6 Inequalities for Geometric Means — 109

- 6.1 Matrix Inequalities for Geometric Means — 109
- 6.2 Symmetric Spaces — 111
- 6.3 Extensions to Lie Groups — 114
- 6.4 Geodesic Triangles in Symmetric Spaces — 114

7 Kostant Convexity Theorems — 121

- 7.1 Kostant Linear Convexity Theorem — 121
- 7.2 A Partial Order — 122
- 7.3 Thompson-Sing and Related Inequalities — 127
- 7.4 Some Matrix Results Associated with $SO(n)$ and $Sp(n)$ — 130
- 7.5 Kostant Nonlinear Convexity Theorem — 133
- 7.6 Thompson Theorem on Complex Symmetric Matrices — 134

Bibliography — 139

Index — 147

Preface

The study of matrix inequalities has a long history and a huge volume of literature can be found in this research area. Conferences, workshops, and mini symposia have been held for its advancement. Many books contain good treatment of matrix inequalities, for example, Bhatia [Bha97, Bha07], Marshall and Olkin and Arnold [MOA11], Horn and Johnson [HJ13, HJ91], Serre [Ser10], Zhan [Zha02], and Zhang [Zha11]. In addition, Bernstein [Ber09] has a large collection of matrix inequalities.

This book is the first attempt to study matrix inequalities and their Lie counterparts in a systematic way. A major inspiration to this work is the seminal paper of Professor B. Kostant [Kos73], in addition to Professor Robert C. Thompson's 1988 Johns Hopkins Lecture Notes, available at http://www.math.sjsu.edu/~so/thompson.html, and many nice matrix inequalities obtained by different authors in the literature. Kostant's paper was motivated by the Schur-Horn Theorem [Sch23, Hor54a] on the eigenvalues and diagonal entries of a Hermitian matrix as well as the Weyl-Horn Theorem [Wey49, Hor54b] on the eigenvalues and singular values of a complex matrix. A good understanding of this paper led the first author [Tam97, Tam99] to discover the connection between the results of Thompson [Tho77] (also see Sing [Sin76]) on the singular values and diagonal entries of a complex matrix. In his review, Professor L. Mirsky praised Thompson's work [Tho77] as "an advance of almost the same order of magnitude as the earlier work of Horn" and, in his review on Sing's paper [Sin76], Thompson commented, "He [Sing] plainly is a talented mathematician from whom many more worthwhile results can be expected."

In his paper [Tho92], Thompson wrote, "Some of the papers in the linear algebra data base achieve their objectives using powerful, advanced tools. Many, however, use only elementary techniques, relying on skill and strategy. These are the high and low roads of the title. Our prediction qualitatively describing the future is that the high-low interaction will yield increasingly deep insight and powerful stimuli. The high road may perhaps be described as 'finding the right ideas' for the correct description of one's problem. It really is quite accurate to state that the young future linear algebraist who hopes to find his own right ideas needs to be trained (at least) in graph theory, Lie theory, functional analysis, multilinear algebra, algebraic geometry, combinatorics, and numerical linear algebra." Though these comments were made 25 years ago, they remain valid and fresh today.

It would be exciting to see whether some beautiful matrix results obtained by the high road approach can be proved by the low road approach, for example, the solution to Horn's conjecture on the spectrum of sum of two Hermitian matrices (see Knutson and Tao [KT99, KT01]) and the convergence of the Aluthge iterations of a complex square matrix (see Antezana and Pujals and Stojanoff [APS11]). Conversely, it would be nice if an insightful explanation, possibly from Lie theory, can be found on Professor Thompson's low road approach in his famous paper [Tho79].

Though Chapter 1 is a review of some materials in matrix theory, it has some special features. Although the additive Jordan decomposition of a matrix has many proofs in the literature, our presentation of Roitman's simple proof [Roi99] is probably the first in a book setting. Moreover, we present Tam's simple induction proof [Tam10a] of A. Horn's result, which is the converse of Weyl's inequalities on the singular values and eigenvalues of a matrix. The Gelfand-Naimark decompostion of a matrix $A = L\omega U$ is given and its difference between and the connection to the Gaussian elimination ωLU are pointed out.

It is our sincere hope that this book is just a beginning of the endeavor of extending matrix inequalities to Lie groups and an inspiration to others.

We are thankful to Callum Fraser, Sarfraz Khan, and Suzanne Lassandro of CRC Press/Taylor & Francis Group for their assistance, support, and patience. Special thanks are given to Luyining (Elaine) Gan for the nice TpX diagrams in this book; Sima Ahsani, Luyining (Elaine) Gan, Wei Gao, Daryl Granario, Mehmet Gumus, Zhuoheng He, Jianzhen (Jason) Liu, Xavier Martinez-Rivera, Samir Raouafi, and Daochang Zhang for proofreading the manuscript (Ph.D. students and visitors of the first author at Auburn University); Beth Fletcher for proofreading the preface and her excellent suggestions. More than likely, any mistakes found in the book are those that we injected after their proofreading.

Finally, we thank our families for their constant support, encouragement, and understanding.

Tin-Yau Tam, Auburn
Xuhua "Roy" Liu, Greenville

February, 2018

Chapter 1

Review of Matrix Theory

1.1	Matrix Decompositions	3
	1.1.1 Polar Decompositions	3
	1.1.2 Singular Value Decomposition	4
	1.1.3 QR Decomposition	4
	1.1.4 Cholesky Decomposition	5
	1.1.5 Additive Decompositions	5
	1.1.6 Jordan Decompositions	6
	1.1.7 LU Decomposition	9
	1.1.8 $L\omega U$ Decomposition	10
1.2	Majorizations	13
1.3	Matrix Norms	19
1.4	The Matrix Exponential Map	23
1.5	Compound Matrices and Applications	25
	1.5.1 Compound Matrices	26
	1.5.2 Additive Compound Matrices	27
	1.5.3 Applications to Matrix Inequalities	29

In this chapter, we review some basic elements of matrix theory that are related to later chapters. The following notations will be used thoroughly.

Let $\mathbb{N}$ denote the set of all positive integers.

Let $\mathbb{Z}$ denote the set of all integers.

Let $\mathbb{R}$ denote the field of all real numbers.

Let $\mathbb{R}_+$ denote the set of all nonnegative real numbers.

Let $\mathbb{C}$ denote the field of all complex numbers.

Let $\mathbb{R}^n$ denote the linear space of all n-tuples over $\mathbb{R}$.

Let $\mathbb{R}^n_+ = \{(x_1, x_2, \ldots, x_n) \in \mathbb{R}^n : x_i \geqslant 0 \text{ for all } 1 \leqslant i \leqslant n\}$.

Let $\mathbb{C}^n$ denote the linear space of all n-tuples over $\mathbb{C}$.

Let $\mathbb{C}_{n \times n}$ denote the linear algebra of all $n \times n$ matrices over $\mathbb{C}$.

Let $\mathbb{R}_{n \times n}$ denote the linear algebra of all $n \times n$ matrices over $\mathbb{R}$.

Let I (or I_n) denote the identity matrix in $\mathbb{C}_{n \times n}$.

For all $A \in \mathbb{C}_{n \times n}$, let $\bar{A}$, $A^\top$, and A^* denote the conjugate, transpose, and conjugate transpose of A, respectively.

For all $A = (a_{ij}) \in \mathbb{C}_{n \times n}$, let

$$s(A) = (s_1(A), \ldots, s_n(A)) = (s_1, \ldots, s_n)$$

denote the vector of *singular values* of A in decreasing order, let

$$\lambda(A) = (\lambda_1(A), \ldots, \lambda_n(A)) = (\lambda_1, \ldots, \lambda_n)$$

denote the vector of *eigenvalues* of A whose absolute values are in decreasing order, let

$$|\lambda(A)| = (|\lambda_1|, \ldots, |\lambda_n|),$$

let

$$(s(A))^m = (s_1^m, \ldots, s_n^m), \qquad \forall\, m \in \mathbb{N},$$

let

$$(\lambda(A))^m = (\lambda_1^m, \ldots, \lambda_n^m), \qquad \forall\, m \in \mathbb{N},$$

let

$$|A| = (A^*A)^{1/2}$$

so that $\lambda(|A|) = s(A)$, let

$$\det A = \prod_{i=1}^n \lambda_i$$

denote the *determinant* of A, and let

$$\operatorname{tr} A = \sum_{i=1}^n a_{ii} = \sum_{i=1}^n \lambda_i$$

denote the *trace* of A.

For scalars $a_1, a_2, \ldots, a_n$, let

$$\operatorname{diag}(a_1, a_2, \ldots, a_n) = \begin{pmatrix} a_1 & & & \\ & a_2 & & \\ & & \ddots & \\ & & & a_n \end{pmatrix}$$

denote the $n \times n$ diagonal matrix whose diagonal entries are $a_1, a_2, \ldots, a_n$, with the prescribed order.

Let $\mathbb{N}_n = \{A \in \mathbb{C}_{n \times n} : A^*A = AA^*\}$ denote the set of all $n \times n$ normal matrices.

Let $\mathbb{H}_n = \{A \in \mathbb{C}_{n \times n} : A^* = A\}$ denote the real linear space of all $n \times n$ Hermitian matrices.

Let $\mathbb{P}_n = \{A \in \mathbb{C}_{n \times n} : x^*Ax > 0 \text{ for all nonzero } x \in \mathbb{C}^n\}$ denote the set of all $n \times n$ positive definite matrices.

Let $\mathbb{S}_n$ denote the group of all $n \times n$ permutation matrices.

Let S_n denote the symmetric group on a finite set of n symbols.

Let $\mathrm{GL}_n(\mathbb{C}) = \{A \in \mathbb{C}_{n \times n} : \det A \neq 0\}$ denote the *general linear group* of all $n \times n$ invertible matrices over $\mathbb{C}$.

Let $\mathrm{GL}_n(\mathbb{R}) = \{A \in \mathbb{R}_{n \times n} : \det A \neq 0\}$ denote the *general linear group* of all $n \times n$ invertible matrices over $\mathbb{R}$.

Let $\mathrm{SL}_n(\mathbb{C}) = \{A \in \mathbb{C}_{n \times n} : \det A = 1\}$ denote the *special linear group* of all $n \times n$ complex matrices whose determinants are 1.

Let $\mathrm{SL}_n(\mathbb{R}) = \{A \in \mathbb{R}_{n \times n} : \det A = 1\}$ denote the *special linear group* of all $n \times n$ real matrices whose determinants are 1.

Let $\mathrm{U}_n = \{U \in \mathbb{C}_{n \times n} : U^*U = UU^* = I\}$ or $\mathrm{U}(n)$ denote the *unitary group* of all $n \times n$ unitary matrices.

Let $\mathrm{O}_n = \{O \in \mathbb{R}_{n \times n} : O^\top O = OO^\top = I\}$ or $\mathrm{O}(n)$ denote the *orthogonal group* of all $n \times n$ real orthogonal matrices.

Let $\mathrm{SU}_n = \{U \in \mathrm{U}_n : \det U = 1\}$ or $\mathrm{SU}(n)$ denote the *special unitary group* of all $n \times n$ unitary matrices whose determinants are 1.

Let $\mathrm{SO}_n = \{O \in \mathrm{O}_n : \det O = 1\}$ or $\mathrm{SO}(n)$ denote the *special orthogonal group* of all $n \times n$ real orthogonal matrices whose determinants are 1.

1.1 Matrix Decompositions

In this section, we review some matrix decompositions that will be used later. They are the left and right polar decompositions, singular value decomposition, QR decomposition, additive and multiplicative Jordan decompositions, and $L\omega U$ decomposition and others. Most of them are familiar to readers with a background of standard matrix analysis (e.g., [HJ13]). Almost all matrix decompositions presented here can be extended to Lie groups or Lie algebras.

1.1.1 Polar Decompositions

For any $A \in \mathbb{C}_{n \times n}$, there exists $U \in \mathrm{U}_n$ such that

$$A = PU, \tag{1.1}$$

where $P = (AA^*)^{1/2}$. Since P is on the left, the decomposition (1.1) is called a *left polar decomposition* of A. Similarly, a *right polar decomposition* for A is

$$A = UP, \tag{1.2}$$

where $P = (A^*A)^{1/2}$ and $U \in U_n$. If A is real, then both P and U in (1.1) and (1.2) may be taken to be real. If $A \in \mathrm{GL}_n(\mathbb{C})$, then $P \in \mathbb{P}_n$ and U is unique in both (1.1) and (1.2).

The polar decompositions (1.1) and (1.2) correspond to the Cartan decompositions $G = PK$ and $G = KP$, respectively, for all connected noncompact real semisimple Lie groups G (see Theorem 2.6 and Example 2.7 for details).

1.1.2 Singular Value Decomposition

For any $A \in \mathbb{C}_{n \times n}$, there exist $U, V \in U_n$ such that

$$A = U \mathrm{diag}\,(s_1, \ldots, s_n) V^*, \tag{1.3}$$

where $s_1 \geqslant s_2 \geqslant \cdots \geqslant s_n \geqslant 0$ are the eigenvalues of $|A| = (A^*A)^{1/2}$, called the *singular values* of A. The columns of U are eigenvectors of AA^* (called *left singular vectors* of A), and the columns of V are eigenvectors of A^*A (called *right singular vectors* of A). However, U and V are never uniquely determined. If $A \in \mathrm{GL}_n(\mathbb{C})$, then $s_n > 0$. The decomposition (1.3) is called a *singular value decomposition* of A, which is equivalent to polar decompositions (1.1) and (1.2).

The singular value decomposition (1.3) corresponds to the Lie group decomposition $G = KA_+K$ for all connected noncompact real semisimple Lie groups G. In such case, the vector of singular values $s(A)$ of $A \in \mathbb{C}_{n \times n}$, in decreasing order, corresponds to the unique A_+-component $a_+(g)$ of $g \in G$ (see Theorem 2.14 and Example 2.15 for details).

1.1.3 *QR* Decomposition

For any $A \in \mathbb{C}_{n \times n}$, there exist $Q \in U_n$ and upper triangular $R \in \mathbb{C}_{n \times n}$ such that

$$A = QR. \tag{1.4}$$

If A is real, then both Q and R may be taken to be real. If $A \in \mathrm{GL}_n(\mathbb{C})$, then the diagonal entries of R may be chosen to be positive; in this case, both Q and R are unique. The decomposition (1.4) is called a *QR decomposition* of A.

The QR decomposition (1.4) is the matrix version of the Gram-Schmidt orthonormalization process on the columns of A from the first column to the last one. When $A \in \mathrm{GL}_n(\mathbb{C})$ and the diagonal entries $r_{ii}, i = 1, \ldots, n$, of R

are positive, r_{ii} is the distance (in Euclidean norm) from the i-th column of A to the subspace spanned by the first $i - 1$ columns of A.

Applications of the Gram-Schmidt process on $A \in \mathbb{C}_{n \times n}$ from the first row to the last row, from the last column to the first column, and from the last row to the first row, respectively, yield the formal decompositions

$$A = LV, \quad A = QL, \quad A = RV,$$

where L and R are lower and upper triangular, respectively, and the unitary V and Q represent that the processes are with respect to rows and columns, respectively.

The QR decomposition (1.4) corresponds to the Iwasawa decomposition $G = KAN$ for all connected noncompact real semisimple Lie groups G. In such case, the diagonal entries of R correspond to the A-component $a(g)$ of $g \in G$ (see Theorem 2.11 and Example 2.12 for details).

1.1.4 Cholesky Decomposition

Any $A \in \mathbb{P}_n$ can be written as

$$A = R^*R, \tag{1.5}$$

where $R \in \mathrm{GL}_n(\mathbb{C})$ is upper triangular with positive diagonal entries. The decomposition (1.5) is called the *Cholesky decomposition*. It can be obtained from the QR decomposition (1.4) as follows: if $A^{1/2} = QR$, then

$$A = A^{1/2} A^{1/2} = (A^{1/2})^* A^{1/2} = (QR)^*(QR) = R^*R.$$

If $S = \{R \in \mathrm{GL}_n(\mathbb{C}) : R \text{ is upper triangular with positive diagonal entries}\}$, then S is a closed solvable subgroup of $\mathrm{GL}_n(\mathbb{C})$. The set $\mathbb{P}_n$ is a closed submanifold of $\mathrm{GL}_n(\mathbb{C})$. The function $\phi : S \to \mathbb{P}_n$ given by $\phi(R) = R^*R$ is actually a diffeomorphism onto $\mathbb{P}_n$. See [Hel78, Proposition VI.5.3] for an extension of this fact to connected noncompact real semisimple Lie groups.

1.1.5 Additive Decompositions

Any $A \in \mathbb{C}_{n \times n}$ can be uniquely written as

$$A = H + S, \tag{1.6}$$

where $H = (A + A^*)/2$ is Hermitian and $S = (A - A^*)/2$ is skew-Hermitian. This is called the *Cartesian decomposition* of A, and it corresponds to the Cartan decomposition $\mathfrak{g} = \mathfrak{k} \oplus \mathfrak{p}$ for all real semisimple Lie algebras $\mathfrak{g}$ (see Theorem 2.3 and Example 2.4 and Example 2.5 for details).

Any $A \in \mathbb{C}_{n \times n}$ can be uniquely written as

$$A = S + D + N, \tag{1.7}$$

where S is skew-Hermitian whose strictly lower triangular entries are the same with A, D is real diagonal (consisting of the real parts of the diagonal entries of A), and N is strictly upper triangular. This corresponds to the Iwasawa decomposition $\mathfrak{g} = \mathfrak{k} \oplus \mathfrak{a} \oplus \mathfrak{n}$ for all real semisimple Lie algebras $\mathfrak{g}$ (see (2.9) and Example 2.10 for details).

1.1.6 Jordan Decompositions

In this subsection, we shall prove an important but less known decomposition, called the *complete multiplicative Jordan decomposition*.

Let $A \in \mathbb{C}_{n \times n}$. Then A is called *nilpotent* if $A^k = 0$ for some $k \in \mathbb{N}$; A is called *semisimple* if it is diagonalizable (that is, if there exists $S \in \mathrm{GL}_n(\mathbb{C})$ such that $S^{-1}AS$ is diagonal); A is called *real semisimple* if it is semisimple and all of its eigenvalues are real.

Let $A \in \mathrm{GL}_n(\mathbb{C})$. Then A is called *elliptic* if it is semisimple and the absolute values of all of its eigenvalues are 1; A is called *hyperbolic* if it is real semisimple and all of its eigenvalues are positive; A is called *unipotent* if $A - I$ is nilpotent. Obviously, each of the three classes of elliptic, hyperbolic, and unipotent matrices is closed under conjugacy (i.e., matrix similarity transformation).

Any $A \in \mathbb{C}_{n \times n}$ can be uniquely written as

$$A = S + N, \tag{1.8}$$

where S is semisimple and N is nilpotent and $SN = NS$. This is called the *additive Jordan decomposition*.

The additive Jordan decomposition (1.8) follows easily from the fact that every $A \in \mathbb{C}_{n \times n}$ is similar to its Jordan normal form J_A (unique up to permutation of the Jordan blocks): If $P^{-1}AP = J_A$ for some $P \in \mathrm{GL}_n(\mathbb{C})$, then write $J_A = J_D + J_N$, where J_D is diagonal and J_N is strictly upper triangular. Note that $J_D J_N = J_N J_D$, since both J_D and J_N are block diagonal matrices with respective blocks of the same size and the blocks of J_D are scalar matrices. Let $S = P^{-1} J_D P$ and $N = P^{-1} J_N P$. Then $A = S + N$ with $SN = NS$. The uniqueness of the additive Jordan decomposition (1.8) follows from the fact that both S and N can be written as polynomials in A without constant terms.

Since other Jordan decompositions in this subsection depend on (1.8), we give a short proof of the Jordan normal form theorem (Theorem 1.1 below is a restatement of the Jordan normal form in the language of linear transformations.)

Let $T : V \to V$ be a linear operator on a finite dimensional vector space V over $\mathbb{C}$. Let $\lambda \in \mathbb{C}$ be an eigenvalue of T. Then there exists a nonzero $x \in V$ such that

$$(T - \lambda I)^k(x) = 0$$

for some $k \in \mathbb{N}$. Let $m \in \mathbb{N}$ be the smallest such integer. The linearly independent set
$$\{(T - \lambda I)^{m-1}(x), \ldots, (T - \lambda I)(x), x\}$$
is called a *Jordan sequence* of T corresponding to λ. The matrix representation of the restriction of T on
$$W = \text{span}\{(T - \lambda I)^{m-1}(x), \ldots, (T - \lambda I)(x), x\}$$
relative to the ordered basis $\{(T-\lambda I)^{m-1}(x), \ldots, (T-\lambda I)(x), x\}$ is the Jordan block
$$J_m(\lambda) = \begin{pmatrix} \lambda & 1 & & \\ & \ddots & \ddots & \\ & & \ddots & 1 \\ & & & \lambda \end{pmatrix}_{m \times m}$$
with $J_1(\lambda) = (\lambda)$. An ordered basis of V consisting of Jordan sequences of T is called a *T-Jordan basis*. Obviously, the matrix representation of T relative to a T-Jordan basis is a Jordan matrix, i.e., a block diagonal matrix
$$J_1 \oplus \cdots \oplus J_s,$$
where the diagonal blocks J_i's, $1 \leqslant i \leqslant s$, are Jordan blocks.

To determining the existence of a Jordan normal form for every $A \in \mathbb{C}_{n \times n}$ is then reduced to the determining the existence of a T-Jordan basis for every linear operator T.

Theorem 1.1. *If $T : V \to V$ is a linear operator on a finite dimensional vector space V over $\mathbb{C}$, then V has a T-Jordan basis.*

Proof. Suppose to the contrary that the statement is false. Let $T : V \to V$ be a counter-example with $\dim V$ minimal. In particular, $V \neq 0$ and $T \neq 0$. Since $\mathbb{C}$ is algebraically closed, the characteristic polynomial of T splits into linear factors. Let μ be an eigenvalue of T. By replacing T by $T - \mu I$ we may assume that $\mu = 0$. So T is not injective and hence $\dim T(V) < \dim V$. By the minimality of $\dim V$, we see that $T(V)$ has a T-Jordan basis, viewing T restricted on $T(V)$. Let $W \supset T(V)$ be a subspace of maximal dimension among all subspaces of V with a T-Jordan basis. We will show that $W = V$, a contradiction.

We first prove that $T(V) = T(W)$. Let β be a T-Jordan basis of W. Because $T(V) \subset W \subset V$, it suffices to show that $\beta \cap T(V) \subset T(W)$. Let $w \in \beta \cap T(V)$. Then w belongs to a Jordan sequence corresponding to some eigenvalue λ of T:
$$\{(T - \lambda I)^{m-1}(x), \ldots, (T - \lambda I)(x), x\}.$$
There are two cases.

Case 1: $\lambda = 0$. If $w \neq x$, then $w = T^i(x) = T(T^{i-1}(x)) \in T(W)$ for some $1 \leqslant i \leqslant m-1$. If $w = x$, pick any $w' \in V$ such that $T(w') = w$. If $w' \notin W$, then we may obtain the following Jordan sequence

$$\{T^{m-1}(x), \ldots, T(x), x, w'\},$$

thus extending β to a T-Jordan basis of a subspace properly containing W, a contradiction. Hence $w' \in W$, and $w \in T(W)$.

Case 2: $\lambda \neq 0$. Then for all $v \in W$ we have

$$v \in T(W) \iff (T - \lambda I)(v) = T(v) - \lambda v \in T(W).$$

Since $(T - \lambda I)^m(w) = 0 \in T(W)$, we see that $w \in T(W)$.

To show $W = V$, let $v \in V$ be arbitrary. Since $T(V) = T(W)$, we have $T(v) = T(w)$ for some $w \in W$. So $v - w \in \ker T$. But $\ker T \subset W$, otherwise we may add to β a nonzero element in $\ker T$ to obtain a T-Jordan basis of a subspace properly containing W. Thus $v \in W$, and so $W = V$. □

Let $A \in \mathrm{GL}_n(\mathbb{C})$. By the additive Jordan decomposition (1.8), we have $A = S + N$ with $S \in \mathrm{GL}_n(\mathbb{C})$. Putting $s = S$ and $u = I + S^{-1}N$, we have the *multiplicative Jordan decomposition*

$$A = su \tag{1.9}$$

where s is semisimple, u is unipotent, and $su = us$. Note that both s and u are unique, by virtue of the uniqueness of S and N in (1.8).

Lemma 1.2. *Any semisimple $s \in \mathrm{GL}_n(\mathbb{C})$ can be uniquely written as*

$$s = eh, \tag{1.10}$$

where e and h are elliptic and hyperbolic, respectively, and $eh = he$.

Proof. Because $s \in \mathrm{GL}_n(\mathbb{C})$ is semisimple, there exist $T \in \mathrm{GL}_n(\mathbb{C})$ and diagonal $D \in \mathrm{GL}_n(\mathbb{C})$ such that $s = TDT^{-1}$. Let E and H be diagonal such that $D = EH$ and that the absolute values of all diagonal entries of E are 1 and that all diagonal entries of H are positive (being the absolute values of the eigenvalues of s). Let $e = TET^{-1}$ and $h = THT^{-1}$ so that e is elliptic and h is hyperbolic. Then $s = eh$ and s, e, h mutually commute. The uniqueness of e and h follows from the uniqueness of the polar decomposition $D = EH$. □

The following result is called *complete multiplicative Jordan decomposition*, abbreviated as CMJD.

Theorem 1.3. *Any $A \in \mathrm{GL}_n(\mathbb{C})$ can be uniquely written as*

$$A = ehu, \tag{1.11}$$

where e, h, and u are elliptic, hyperbolic, and unipotent, respectively, and all three commute.

Proof. Since $A \in \mathrm{GL}_n(\mathbb{C})$, by (1.9) and (1.10), there exist unique semisimple s, elliptic e, hyperbolic h, and unipotent u such that

$$A = su = ehu$$

with $su = us$ and $eh = he$. Since $ehu = A = ueh = (ueu^{-1})(uhu^{-1})u$ with elliptic ueu^{-1} and hyperbolic uhu^{-1}, the uniqueness of s, e, h, and u implies that

$$e = ueu^{-1} \quad \text{and} \quad h = uhu^{-1}.$$

Therefore, $ue = eu$ and $uh = hu$. □

We emphasize that in the decomposition (1.11), the eigenvalues of h are the absolute values of the corresponding eigenvalues of A, i.e.,

$$\lambda(h) = |\lambda(A)|.$$

We also note that if $A \in \mathrm{GL}_n(\mathbb{R})$ and $A = ehu$ is the complete multiplicative Jordan decomposition of A viewed as for $A \in \mathrm{GL}_n(\mathbb{C})$, then e, h, u are all real. To see this, we first recall that the real matrix A is similar via a *real* similarity matrix to its *real Jordan canonical form* [HJ13, p.202]. It follows that the matrices in the additive Jordan decomposition $A = S + N$ and the multiplicative Jordan decomposition $A = su$ are real. Furthermore, for $a, b \in \mathbb{R}$, the polar decomposition

$$\begin{pmatrix} a & b \\ -b & a \end{pmatrix} = \begin{pmatrix} \frac{a}{\sqrt{a^2+b^2}} & \frac{b}{\sqrt{a^2+b^2}} \\ -\frac{b}{\sqrt{a^2+b^2}} & \frac{a}{\sqrt{a^2+b^2}} \end{pmatrix} \begin{pmatrix} \sqrt{a^2+b^2} & 0 \\ 0 & \sqrt{a^2+b^2} \end{pmatrix}$$

guarantees that matrices e and h in the unique decomposition $s = eh$ are also real.

The complete multiplicative Jordan decomposition (1.10) has an extension, bearing the same name, to all connected noncompact real semisimple Lie groups (see Theorem 2.17).

1.1.7 *LU* Decomposition

Another famous matrix decomposition is the *LU* decomposition, which can be viewed as the matrix form of Gaussian elimination. Though we will not use the *LU* decomposition in our study, we state it here for comparison with a similar matrix decomposition that is extendable to Lie groups.

For $A \in \mathbb{C}_{n \times n}$, if there exist a lower triangular $L \in \mathbb{C}_{n \times n}$ and an upper triangular $U \in \mathbb{C}_{n \times n}$ such that

$$A = LU, \tag{1.12}$$

then such a decomposition is called an *LU decomposition* of A. Not every square matrix has an LU decomposition. If $A \in \mathrm{GL}_n(\mathbb{C})$, then A has an LU decomposition if and only if every leading principal submatrix of A is invertible; in this case, A may be uniquely written as

$$A = L'DU', \tag{1.13}$$

where $L' \in \mathrm{GL}_n(\mathbb{C})$ is unit lower triangular, $U' \in \mathrm{GL}_n(\mathbb{C})$ is unit upper triangular, and $D \in \mathrm{GL}_n(\mathbb{C})$ is a diagonal matrix whose leading principal minors are equal to the corresponding ones of A.

If $Ax = b$ is a linear system with $A \in \mathbb{C}_{n \times n}$, one can always reorder the equations so that the new coefficient matrix has an LU decomposition. In other words, every $A \in \mathbb{C}_{n \times n}$ can be written as

$$A = \omega LU, \tag{1.14}$$

where ω is a permutation matrix.

1.1.8 $L\omega U$ Decomposition

Let $\mathbb{F} = \mathbb{R}$ or $\mathbb{C}$. The $L\omega U$ decomposition asserts that each $A \in \mathrm{GL}_n(\mathbb{F})$ can be written as

$$A = L\omega U, \tag{1.15}$$

where $L \in \mathrm{GL}_n(\mathbb{F})$ is unit lower triangular, $U \in \mathrm{GL}_n(\mathbb{F})$ is upper triangular, and ω is a permutation matrix. It is different from the ωLU decomposition (1.14). Though the $L\omega U$ decomposition is less known, it has very nice properties. For example, the permutation matrix ω and $\mathrm{diag}\, U \in \mathbb{F}^n$ in (1.15) are uniquely determined by A, while none of the components in (1.14) is unique.

Let $\{e_1, \ldots, e_n\}$ be the standard basis of $\mathbb{F}^n$, that is, e_i has 1 as the only nonzero entry at the i-th entry. Using the same notation, we identify a permutation $\omega \in S_n$ with the unique permutation matrix $\omega \in \mathbb{S}_n$, where $\omega e_i = e_{\omega(i)}$. The matrix representation of $\omega \in S_n$ under the standard basis is

$$\omega = \left(e_{\omega(1)}, \ldots, e_{\omega(n)}\right).$$

Write $A = (a_1, \ldots, a_n)$ in column form. Then $A\omega = \left(a_{\omega(1)}, \ldots, a_{\omega(n)}\right)$. Moreover, if $x_1, \ldots, x_n \in \mathbb{F}$, then

$$\omega^{-1} \mathrm{diag}\,(x_1, \ldots, x_n) \omega = \mathrm{diag}\,(x_{\omega(1)}, \ldots, x_{\omega(n)}). \tag{1.16}$$

Given a matrix $A \in \mathbb{F}_{n \times n}$, let $A(i|j)$ denote the submatrix formed by the first i rows and the first j columns of A, where $1 \leqslant i,j \leqslant n$. The following result is the $L\omega U$ decomposition, also known as the *Gelfand-Naimark decomposition*.

Theorem 1.4. *Let $\mathbb{F} = \mathbb{R}$ or $\mathbb{C}$. If $A \in \mathrm{GL}_n(\mathbb{F})$, then there exist a permutation* matrix $\omega \in \mathbb{S}_n$, a unit lower triangular matrix $L \in \mathrm{GL}_n(\mathbb{F})$, and an upper triangular $U \in \mathrm{GL}_n(\mathbb{F})$ such that

$$A = L\omega U.$$

The permutation matrix ω is uniquely determined by A:

$$\mathrm{rank}\,\omega(i|j) = \mathrm{rank}\,A(i|j), \qquad \forall\, 1 \leqslant i, j \leqslant n.$$

Moreover, $\mathrm{diag}\,U$ is uniquely determined by A.

Proof. We first prove the existence of the decomposition $A = L\omega U$, which is a matrix version of some sequence of elementary row and column operations applied to A.

Suppose that a_{k1}, $1 \leqslant k \leqslant n$, is the first nonzero entry of the first column of A. By multiplying the first column of A by $1/a_{k1}$, we turn the $(k, 1)$ entry to 1. Using this 1 as a pivot, we consecutively eliminate other nonzero entries on the first column (using row operations) and the kth row (using column operations).

The above operations are equivalent to the following post- and pre-matrix multiplications: Let $D_1 = \mathrm{diag}\,(1/a_{k1}, 1, \ldots, 1) \in \mathrm{GL}_n(\mathbb{F})$. Let $A' = AD_1$ and denote the (i, j) entry of A' by a'_{ij}. Let $E_{ij} \in \mathbb{R}_{n \times n}$ with (i, j) entry 1 as the only nonzero entry. Let

$$L_1 = (I - a'_{k+1,1}E_{k+1,k})(I - a'_{k+2,1}E_{k+2,k}) \cdots (I - a'_{n1}E_{nk}) \in \mathrm{GL}_n(\mathbb{F}),$$

a unit lower triangular matrix, and

$$U_1 = (I - a'_{k2}E_{12})(I - a'_{k3}E_{13}) \cdots (I - a'_{kn}E_{1n}) \in \mathrm{GL}_n(\mathbb{F}),$$

a unit upper triangular matrix. Then

$$
\begin{array}{ccccc}
A & \to & L_1 A D_1 & \to & L_1 A D_1 U_1 \\
\end{array}
$$

$$
\begin{pmatrix}
0 & * & \cdots & * & * & * & \cdots & * \\
\vdots & \vdots & & \vdots & \vdots & \vdots & & \vdots \\
0 & * & \cdots & * & * & * & \cdots & * \\
a_{k1} & * & \cdots & * & * & * & \cdots & * \\
* & * & \cdots & * & * & * & \cdots & * \\
\vdots & \vdots & & \vdots & \vdots & \vdots & & \vdots \\
* & * & \cdots & * & * & * & \cdots & *
\end{pmatrix}
\to
\begin{pmatrix}
0 & * & \cdots & * & * & * & \cdots & * \\
\vdots & \vdots & & \vdots & \vdots & \vdots & & \vdots \\
0 & * & \cdots & * & * & * & \cdots & * \\
1 & * & \cdots & * & * & * & \cdots & * \\
0 & * & \cdots & * & * & * & \cdots & * \\
\vdots & \vdots & & \vdots & \vdots & \vdots & & \vdots \\
0 & * & \cdots & * & * & * & \cdots & *
\end{pmatrix}
\to
\begin{pmatrix}
0 & * & \cdots & * & * & * & \cdots & * \\
\vdots & \vdots & & \vdots & \vdots & \vdots & & \vdots \\
0 & * & \cdots & * & * & * & \cdots & * \\
1 & 0 & \cdots & 0 & 0 & 0 & \cdots & 0 \\
0 & * & \cdots & * & * & * & \cdots & * \\
\vdots & \vdots & & \vdots & \vdots & \vdots & & \vdots \\
0 & * & \cdots & * & * & * & \cdots & *
\end{pmatrix}.
$$

Repeat the procedure on the second column of $L_1 A D_1 U_1$, and so on. Eventually we obtain a permutation matrix ω, unit lower triangular matrices $L_1, \ldots, L_n \in \mathrm{GL}_n(\mathbb{F})$, diagonal matrices $D_1, \ldots, D_n \in \mathrm{GL}_n(\mathbb{F})$, and unit upper triangular matrices $U_1, \ldots, U_n \in \mathrm{GL}_n(\mathbb{F})$ such that

$$L_n \cdots L_1 A D_1 U_1 \cdots D_n U_n = \omega.$$

Putting

$$L^{-1} = L_n \cdots L_1,$$
$$U^{-1} = D_1 U_1 \cdots D_n U_n,$$

we obtain the desired decomposition $A = L\omega U$.

Since the group of nonsingular diagonal matrices normalizes the group of unit upper triangular matrices, we have

$$U^{-1} = U'D,$$

where

$$U' = (D_1 U_1 D_1^{-1})(D_1 D_2 U_2 (D_1 D_2)^{-1}) \cdots (D_1 \cdots D_n U_n (D_1 \cdots D_n)^{-1})$$

is unit upper triangular and $D = D_1 \cdots D_n$. So $U = D^{-1}U'^{-1}$. In other words, the ith diagonal entry u_{ii} of U is the first nonzero entry of the ith column of $L_{i-1} \cdots L_1 A D_1 U_1 \cdots D_{i-1} U_{i-1}$, according to the definition of D_i in the ith elimination step.

By block multiplication we notice that

$$A(i|j) = \begin{pmatrix} L(i|i) & 0 \end{pmatrix} \begin{pmatrix} \omega(i|j) & * \\ * & * \end{pmatrix} \begin{pmatrix} U(j|j) \\ 0 \end{pmatrix}$$
$$= L(i|i)\omega(i|j)U(j|j)$$

where $A(i|j)$ denotes the submatrix of A obtained by deleting the ith row and jth column of A. Thus

$$\operatorname{rank}\omega(i|j) = \operatorname{rank} A(i|j), \qquad \forall 1 \leqslant i, j \leqslant n.$$

Obviously $\operatorname{rank}\omega(i|j)$ is the number of nonzero entries in $\omega(i|j)$. Thus it is easy to verify that ω_{ij} is nonzero if and only if

$$\operatorname{rank}\omega(i|j) - \operatorname{rank}\omega(i|j-1) - \operatorname{rank}\omega(i-1|j) + \operatorname{rank}\omega(i-1|j-1) = 1.$$

So the permutation matrix ω is uniquely determined by

$$\operatorname{rank}\omega(i|j), \qquad \forall 1 \leqslant i, j \leqslant n.$$

Hence ω is uniquely determined by A.

If $L\omega U = L'\omega U'$ for another unit lower triangular L' and upper triangular U', then $\omega^{-1}L'^{-1}L\omega = U'U^{-1}$. Clearly the diagonal entries of $\omega^{-1}L'^{-1}L\omega$ are 1's, so $\operatorname{diag} U = \operatorname{diag} U'$. Thus $\operatorname{diag} U$ is uniquely determined by A. □

Now the following is well-defined:

$$u(A) := \operatorname{diag} U = \operatorname{diag}(u_{11}, \ldots, u_{nn}), \tag{1.17}$$

where $A = L\omega U$ is any Gelfand-Naimark decomposition of A.

Although ω and $u(A)$ are unique in all Gelfand-Naimark decompositions $A = L\omega U$ of A, the components L and U may be not unique. For example,

$$\begin{pmatrix} 0 & 1 \\ 1 & 1 \end{pmatrix} = \begin{pmatrix} 1 & 0 \\ 1 & 1 \end{pmatrix} \begin{pmatrix} 0 & 1 \\ 1 & 0 \end{pmatrix} \begin{pmatrix} 1 & 0 \\ 0 & 1 \end{pmatrix} = \begin{pmatrix} 1 & 0 \\ 0 & 1 \end{pmatrix} \begin{pmatrix} 0 & 1 \\ 1 & 0 \end{pmatrix} \begin{pmatrix} 1 & 1 \\ 0 & 1 \end{pmatrix}.$$

The unique ω in a Gelfand-Naimark decomposition $A = L\omega U$ of A can also be the permutation matrix in some Gaussian elimination $A = \omega L'U'$ of A. To see this, notice that $\omega^{-1}A = (\omega^{-1}L\omega)U$ and

$$\det\left[(\omega^{-1}L\omega)(k|k)\right] = 1,$$

since $(\omega^{-1}L\omega)(k|k)$ is the submatrix formed by choosing the $\omega(1), \cdots, \omega(k)$ rows and columns of L. Therefore, by (1.13) we have $\omega^{-1}L\omega = L_1 U_1$ for some unit lower triangular L_1 and unit upper triangular U_1, and

$$A = L\omega U = \omega(\omega^{-1}L\omega)U = \omega L_1(U_1 U) = \omega L'U', \qquad (1.18)$$

where $L' = L_1$ and $U' = U_1 U$.

Furthermore, by (1.18) we have $\omega^{-1}A = L_1 U_1 U$. Hence $u(A)$ can be computed by

$$\prod_{i=1}^{k} u_{ii} = \det\left[U(k|k)\right] = \det\left[(L_1 U_1 U)(k|k)\right] = \det\left[(\omega^{-1}A)(k|k)\right]. \qquad (1.19)$$

Let $J \in \mathrm{GL}_n(\mathbb{C})$ denote the permutation matrix whose anti-diagonal entries are 1. Note that $J^2 = I$, JLJ is upper triangular, and JUJ is lower triangular. We then get two variations of $L\omega U$ decomposition: If $AJ = L\omega U$, then

$$A = L(\omega J)(JUJ) = L\omega' L'. \qquad (1.20)$$

If $JA = L\omega U$, then

$$A = (JLJ)(J\omega)U = U'\omega' U \qquad (1.21)$$

which is cited as a result by Gelfand-Naimark in [Hel78, p.434]. The Gelfand-Naimark decomposition and its variations (1.20) and (1.21) correspond to variations of the Bruhat decomposition of connected noncompact real semisimple Lie groups (see [Lia04, p.117]).

Notes and References. The proof of Theorem 1.1 is adopted from [Roi99]. The proof of Theorem 1.4 is from [HT10].

1.2 Majorizations

A *preorder* on a set is a binary relation that is reflexive and transitive. If a preorder is also anti-symmetric, then it is called a *partial order*. A natural partial order on $\mathbb{R}^n$ is defined by the cone $\mathbb{R}_+^n$:

$$x \leqslant y \quad \Longleftrightarrow \quad y - x \in \mathbb{R}_+^n.$$

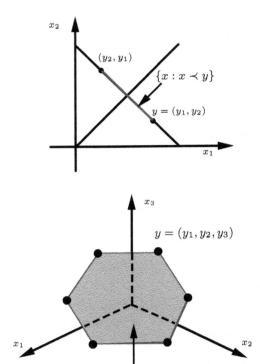

Let $x = (x_1, x_2, \ldots, x_n)$ and $y = (y_1, y_2, \ldots, y_n)$ be in $\mathbb{R}^n$. Let $x^{\downarrow} = (x_{[1]}, x_{[2]}, \ldots, x_{[n]})$ denote a rearrangement of the components of x such that $x_{[1]} \geqslant x_{[2]} \geqslant \cdots \geqslant x_{[n]}$. We say that x is *weakly majorized* by y, denoted by $x \prec_{\mathrm{w}} y$, if

$$\sum_{i=1}^{k} x_{[i]} \leqslant \sum_{i=1}^{k} y_{[i]}, \quad \forall\, 1 \leqslant k \leqslant n.$$

If $x \prec_{\mathrm{w}} y$ and $\sum_{i=1}^{n} x_{[i]} = \sum_{i=1}^{n} y_{[i]}$, then we say that x is *majorized* by y and denote it by

$$x \prec y.$$

The above diagrams are majorization regions when $n = 2$ and $n = 3$, respectively.

When x and y are nonnegative, we say that x is *weakly log-majorized* by y, denoted by $x \prec_{\mathrm{w\text{-}log}} y$, if

$$\prod_{i=1}^{k} x_{[i]} \leqslant \prod_{i=1}^{k} y_{[i]}, \quad \forall\, 1 \leqslant k \leqslant n.$$

If $x \prec_{\text{w-log}} y$ and $\prod_{i=1}^{n} x_{[i]} = \prod_{i=1}^{n} y_{[i]}$, then we say that x is *log-majorized* by y and denote it by

$$x \prec_{\log} y.$$

In other words, when x and y are positive,

$$x \prec_{\log} y \iff \log x \prec \log y.$$

Each of the above four types of relation is a preorder (but not a partial order) on $\mathbb{R}^n$ or $\mathbb{R}^n_+$.

There are many equivalent conditions for majorization $\prec$ and one of them is suitable for generalization to Lie groups and Lie algebras. We define some terms first.

The *convex hull* of a finite point set S in $\mathbb{R}^n$ is the set of all convex combinations of its points. Denote the convex hull of $S = \{p_1, p_2, \ldots, p_m\} \subset \mathbb{R}^n$ by conv S, i.e.,

$$\text{conv } S = \left\{ \sum_{i=1}^{m} c_i p_i : 0 \leqslant c_i \leqslant 1 \text{ for all } 1 \leqslant i \leqslant m \text{ and } \sum_{i=1}^{m} c_i = 1 \right\}.$$

Geometrically, conv S is the convex polytope in $\mathbb{R}^n$ whose vertices are the points of S.

An entry-wise nonnegative matrix $A \in \mathbb{R}_{n \times n}$ is said to be *doubly stochastic* if all its row sums and column sums are 1. The following famous characterization of doubly stochastic matrices is due to Birkhoff (see [HJ13, p.549]).

Theorem 1.5. (Birkhoff) *A matrix is doubly stochastic if and only if it is a convex combination of permutation matrices.*

The following is a collection of equivalent conditions for majorization. The equivalence (1) $\Leftrightarrow$ (4) can be extended to real semisimple Lie groups and Lie algebras, where the symmetric group S_n is replaced by the Weyl group (see Example 2.22).

Theorem 1.6. *The following statements are equivalent for $x, y \in \mathbb{R}^n$.*

(1) $x \prec y$.

(2) $x = Ay$ *for some doubly stochastic matrix* $A \in \mathbb{R}_{n \times n}$.

(3) $x \in \text{conv } S_n \cdot y$, *where* $S_n \cdot y$ *denotes the orbit of* y *under the action of the symmetric group* S_n.

(4) $\text{conv } S_n \cdot x \subset \text{conv } S_n \cdot y$.

Proof. We shall show that $(1) \Leftrightarrow (2) \Leftrightarrow (3) \Leftrightarrow (4)$.

$(1) \Leftrightarrow (2)$. Note that the majorization relationship is invariant under rearrangement of the entries of x and y. The product of doubly stochastic matrices is also doubly stochastic. Thus we may assume without loss of generality that $x = (x_1, x_2, \ldots, x_n)$ and $y = (y_1, y_2, \ldots, y_n)$ with $x_1 \geqslant x_2 \geqslant \cdots \geqslant x_n$ and $y_1 \geqslant y_2 \geqslant \cdots \geqslant y_n$.

For $(1) \Rightarrow (2)$, we proceed by induction on n. The case $n = 1$ is trivial, so we consider $n \geqslant 2$. Since $x \prec y$, it follows that

$$y_1 \geqslant x_1 \geqslant y_k$$

for some minimal $1 \leqslant k \leqslant n$. There are two cases under consideration: $k = 1$ and $k \geqslant 2$.

Case 1: $k = 1$. Then $x_1 = y_1$ and $(x_2, \ldots, x_n) := a \prec b =: (y_2, \ldots, y_n)$. By the induction hypothesis, there exists an $(n-1) \times (n-1)$ doubly stochastic matrix X such that $a = Xb$. So $x = Ay$ for doubly stochastic

$$A = \begin{pmatrix} 1 & 0 \\ 0 & X \end{pmatrix}.$$

Case 2: $k \geqslant 2$. Since $x_1 < y_1$, there exist $t \in [0, 1)$ such that

$$x_1 = ty_1 + (1-t)y_k.$$

Now $Yy = (x_1, y_2, \ldots, y_{k-1}, ty_k + (1-t)y_1, y_{k+1}, y_n)$, where

$$Y = \begin{pmatrix} t & & 1-t & \\ & I_{k-2} & & \\ 1-t & & t & \\ & & & I_{n-k} \end{pmatrix}$$

is doubly stochastic. Put $c := (x_2, \ldots, x_n)$ and $d := (y_2, \ldots, y_{k-1}, ty_k + (1-t)y_1, y_{k+1}, y_n)$. Since $x_n \leqslant \cdots x_2 \leqslant x_1 \leqslant y_{k-1} \leqslant \cdots \leqslant y_2$, we have for all $2 \leqslant l \leqslant k-1$

$$\sum_{i=2}^{l} x_i \leqslant \sum_{i=2}^{l} y_i.$$

Now for $k \leqslant l \leqslant n$, we have

$$\sum_{i=2}^{l} x_i = \sum_{i=1}^{l} x_i - x_1 \leqslant \sum_{i=1}^{l} y_i - (ty_1 + (1-t)y_k) = \sum_{i=2}^{k-1} y_i + (ty_k + (1-t)y_1) + \sum_{i=k+1}^{l} y_i.$$

It follows that $c \prec d$. By the induction hypothesis, there exists an $(n-1) \times (n-1)$ doubly stochastic matrix Z such that $c = Zd$. Now

$$x = \begin{pmatrix} x_1 \\ c \end{pmatrix} = \begin{pmatrix} 1 & 0 \\ 0 & Z \end{pmatrix} \begin{pmatrix} x_1 \\ d \end{pmatrix} = \begin{pmatrix} 1 & 0 \\ 0 & Z \end{pmatrix} Yy.$$

Since the product of two doubly stochastic matrices is doubly stochastic, $A = (1 \oplus Z)Y$ is doubly stochastic, and $x = Ay$.

For (2) $\Rightarrow$ (1), suppose $x = Ay$ for some doubly stochastic $A = (a_{ij}) \in \mathbb{R}_{n \times n}$. For each $1 \leqslant k \leqslant n-1$, let

$$C_{k,j} = \sum_{i=1}^{k} a_{ij}$$

be the sum of the first k entries of the j-th column. Because A is doubly stochastic, we have $0 \leqslant C_{k,j} \leqslant 1$ and $\sum_{j=1}^{n} C_{k,j} = k$. For $1 \leqslant k \leqslant n-1$, we have

$$\sum_{i=1}^{k} x_i = \sum_{i=1}^{k}\sum_{j=1}^{n} a_{ij}y_j = \sum_{j=1}^{n} C_{k,j}y_j$$
$$\leqslant \sum_{j=1}^{k-1} C_{k,j}y_j + \sum_{j=k}^{n} C_{k,j}y_k$$
$$= \sum_{j=1}^{k-1} C_{k,j}y_j + (k - \sum_{j=1}^{k-1} C_{k,j})y_k$$
$$= \sum_{j=1}^{k-1} C_{k,j}(y_j - y_k) + ky_k$$
$$\leqslant \sum_{j=1}^{k-1} (y_j - y_k) + ky_k$$
$$= \sum_{j=1}^{k} y_j.$$

Let $e = (1, 1, \ldots, 1) \in \mathbb{R}^n$. Then $A^\top e = e$ and

$$\sum_{i=1}^{n} x_i = \langle x, e \rangle = \langle Ay, e \rangle = \langle y, A^\top e \rangle = \langle y, e \rangle = \sum_{i=1}^{n} y_i,$$

where $\langle u, v \rangle = v^\top u$ denotes the usual inner product on $\mathbb{R}^n$. So $x \prec y$.

(2) $\Leftrightarrow$ (3). This follows from the Birkhoff Theorem (Theorem 1.5).

(3) $\Leftrightarrow$ (4). We only show (3) $\Rightarrow$ (4), for the converse is trivial. Suppose $x \in \operatorname{conv} S_n \cdot y$. Then there exist permutation matrices $P_1, P_2, \ldots, P_m \in \mathbb{S}_n$ and nonnegative numbers $c_1, c_2, \ldots, c_m \in \mathbb{R}$ such that

$$x = c_1(P_1 y) + c_2(P_2 y) + \cdots + c_m(P_m y) \quad \text{and} \quad \sum_{i=1}^{m} c_i = 1.$$

For any choice of permutation matrices $Q_1, Q_2, \ldots, Q_l \in \mathbb{S}_n$ and nonnegative numbers $d_1, d_2, \ldots, d_l \in \mathbb{R}$ such that $\sum_{j=1}^{l} d_j = 1$, we have

$$\sum_{j=1}^{l} d_j(Q_j x) = \sum_{j=1}^{l} d_j Q_j \sum_{i=1}^{m} c_i(P_i y) = \sum_{j=1}^{l} \sum_{i=1}^{m} a_{(j,i)} R_{(j,i)} y$$

with $R_{(j,i)} = Q_j P_i \in \mathbb{S}_n$ and $a_{(j,i)} = d_j c_i \geqslant 0$ and

$$\sum_{j=1}^{l} \sum_{i=1}^{m} a_{(j,i)} = \sum_{j=1}^{l} \sum_{i=1}^{m} d_j c_i = \left(\sum_{j=1}^{l} d_j\right)\left(\sum_{i=1}^{m} c_i\right) = 1.$$

This shows that conv $\mathbb{S}_n \cdot x \subset$ conv $\mathbb{S}_n \cdot y$. $\square$

The following result, as a special case of [Wey49, Lemma], is useful in later chapters.

Theorem 1.7. (Weyl) *If $x = (x_1, \ldots, x_n) \in \mathbb{R}_+^n$ and $y = (y_1, \ldots, y_n) \in \mathbb{R}_+^n$, then*

$$x \prec_{w\text{-log}} y \quad \Rightarrow \quad x \prec_w y.$$

In particular, $x \prec_{\log} y$ implies $x \prec_w y$.

We conclude this section with the following remark.

Remark 1.8. There are two important orderings that can be defined on the real linear space $\mathbb{H}_n$ of all $n \times n$ Hermitian matrices:

(1) the Löwner order (a partial order) defined by

$$A \leqslant B \quad \Longleftrightarrow \quad B - A \text{ is positive semidefinite};$$

(2) the majorization (a preorder) defined by

$$A \prec B \quad \Longleftrightarrow \quad \lambda(A) \prec \lambda(B).$$

For $A \neq B$ in $\mathbb{H}_n$, the two relations $A \leqslant B$ and $A \prec B$ never coincide. This is because

$$A \leqslant B \quad \Rightarrow \quad \lambda_i(A) \leqslant \lambda_i(B), \quad \forall i = 1, \ldots, n.$$

Notes and References. Majorizations are extensively studied in the celebrated monograph [MOA11]. In this work, we are interested in matrix inequalities in the form of majorization and log-majorization, as well as their extensions in Lie groups and Lie algebras.

1.3 Matrix Norms

Let $\mathbb{F}$ be the field $\mathbb{C}$ or $\mathbb{R}$. Let V be a finite dimensional vector space over $\mathbb{F}$. A *norm* on V is a function $\|\cdot\| : V \to \mathbb{R}$ satisfying the following properties:

(1) $\|v\| \geqslant 0$ for all $v \in V$ with equality if and only if $v = 0$.

(2) $\|\alpha v\| = |\alpha| \|v\|$ for all $\alpha \in \mathbb{F}$ and $v \in V$.

(3) $\|v + w\| \leqslant \|v\| + \|w\|$ for all $v, w \in V$.

The *Euclidean norm* $\|\cdot\|_2$ on $\mathbb{C}^n$ is defined by

$$\|x\|_2 = \sqrt{\sum_{i=1}^{n} |x_i|^2}, \tag{1.22}$$

for all $x = (x_1, x_2, \ldots, x_n) \in \mathbb{C}^n$.

A norm on $\mathbb{C}_{m \times n}$ is a *vector norm on matrices*. If $\|\cdot\| : \mathbb{C}_{n \times n} \to \mathbb{R}$ is a vector norm and

$$\|AB\| \leqslant \|A\| \, \|B\|$$

for all $A, B \in \mathbb{C}_{n \times n}$, then $\|\cdot\|$ is called a *matrix norm*.

The *Frobenius norm* $\|\cdot\|_F$ on $\mathbb{C}_{n \times n}$ is defined by

$$\|A\|_F = \sqrt{\sum_{i,j=1}^{n} |a_{ij}|^2} = \sqrt{\operatorname{tr} A^* A} = \sqrt{\sum_{i=1}^{n} [s_i(A)]^2}, \tag{1.23}$$

for all $A = (a_{ij}) \in \mathbb{C}_{n \times n}$.

The *spectral norm* $\|\cdot\|$ on $\mathbb{C}_{n \times n}$ is defined by

$$\|A\| = \max_{\substack{x \in \mathbb{C}^n \\ \|x\|_2 = 1}} \|Ax\|_2 = s_1(A) \tag{1.24}$$

for all $A \in \mathbb{C}_{n \times n}$.

A norm $\|\|\cdot\|\|$ on $\mathbb{C}_{n \times n}$ is said to be *unitarily invariant* if

$$\|\|UAV\|\| = \|\|A\|\|$$

for all $A \in \mathbb{C}_{n \times n}$ and for all $U, V \in \mathbf{U}_n$. It is shown in [HJ13, p.469] that a unitarily invariant norm $\|\|\cdot\|\|$ on $\mathbb{C}_{n \times n}$ is a matrix norm if and only if $\|\|A\|\| \geqslant s_1(A)$ for all $A \in \mathbb{C}_{n \times n}$.

Both the Frobenius norm and the spectral norm are unitarily invariant matrix norms. In fact, they are special cases of the class of *Schatten p-norms* on $\mathbb{C}_{n\times n}$ defined by

$$\|A\|_p = \left(\operatorname{tr}(A^*A)^{p/2}\right)^{1/p} = \left(\sum_{i=1}^n [s_i(A)]^p\right)^{1/p}, \quad \forall\, 1 \leqslant p < \infty. \qquad (1.25)$$

Because $\lim_{p\to\infty} \|A\|_p = s_1(A)$, it is natural to define

$$\|A\|_\infty = s_1(A).$$

By singular value decomposition, it is easy to see that all Schatten p-norms are unitarily invariant matrix norms.

Another important class of unitarily invariant matrix norms on $\mathbb{C}_{n\times n}$ is the *Ky Fan k-norm* defined by

$$\|A\|_{(k)} = \sum_{i=1}^k s_i(A), \quad \forall\, 1 \leqslant k \leqslant n. \qquad (1.26)$$

Ky Fan k-norms are important because of the *Fan Dominance Theorem* (Theorem 1.11), which follows from a property of symmetric gauge functions and a characterization of unitarily invariant norm in terms of symmetric gauge functions.

A function $\Phi : \mathbb{R}^n \to \mathbb{R}$ is called a *symmetric gauge function* if

(1) Φ is a norm,

(2) $\Phi(Px) = \Phi(x)$ for all $P \in \mathbb{S}_n$ and $x \in \mathbb{R}^n$, and

(3) $\Phi(\varepsilon_1 x_1, \ldots, \varepsilon_n x_n) = \Phi(x_1, \ldots, x_n)$ if $\varepsilon_i = \pm 1$ for all $1 \leqslant i \leqslant n$.

Because of (3), a symmetric gauge function is completely determined by its values on $\mathbb{R}^n_+$. Symmetric gauge functions on $\mathbb{R}^n$ are also called *symmetric gauge invariant norms*.

The following result is a characterization of weak-majorization in terms of symmetric gauge functions given by Ky Fan [Fan51, Theorem 4].

Theorem 1.9. (Fan) *Let $x, y \in \mathbb{R}^n$. Then*

$$x \prec_w y$$

if and only if

$$\Phi(x) \leqslant \Phi(y)$$

for every symmetric gauge function Φ on $\mathbb{R}^n$.

Proof. We assume that $x = (x_1, \ldots, x_n)$ with $x_1 \geqslant \cdots \geqslant x_n$ and $y = (y_1, \ldots, y_n)$ with $y_1 \geqslant \cdots \geqslant y_n$, since both symmetric gauge functions and weak majorization relations are invariant under permutation.

($\Leftarrow$). Consider the special symmetric gauge functions Φ_k, for $1 \leqslant k \leqslant n$, defined by

$$\Phi_k(z_1, \ldots, z_n) = \max_{1 \leqslant i_1 \leqslant \cdots \leqslant i_k \leqslant n} \sum_{j=1}^{k} |z_{i_j}|.$$

Then $\Phi_k(x) \leqslant \Phi_k(y)$ for all $1 \leqslant k \leqslant n$ means that $x \prec_w y$.

($\Rightarrow$). Let Φ be any symmetric gauge function Φ on $\mathbb{R}^n$. Since $x \prec_w y$, by adjusting y_n and keeping the other components of y, we can obtain some $y' \in \mathbb{R}^n$ such that

$$x \prec y' \leqslant y.$$

By Theorem 1.6, there exists a doubly stochastic matrix A such that $x = Ay'$. It follows from Theorem 1.6 that

$$x = Ay' \leqslant Ay = \left(\sum_{i=1}^{m} c_i P_i \right) y \prec y$$

for some $P_1, \ldots, P_m \in \mathbb{S}_n$ and $c_1, \ldots, c_m \in [0,1]$ such that $c_1 + \cdots + c_m = 1$. Note that Φ is increasing on $\mathbb{R}_+^n$ with respect to the partial order $\leqslant$, which follows from the fact that Φ obeys the triangle inequality, since it is a norm. Therefore, we have

$$\Phi(x) \leqslant \Phi(Ay) \leqslant \sum_{i=1}^{m} c_i \Phi(P_i y) = \sum_{i=1}^{m} c_i y = y.$$

This completes the proof. $\square$

The following result is a characterization of unitarily invariant norms in terms of symmetric gauge functions given by von Neumann [vN37].

Theorem 1.10. (von Neumann) *If $||| \cdot |||$ is a unitarily invariant norm on $\mathbb{C}_{n \times n}$, then the function on $\mathbb{R}^n$ defined by*

$$\Phi_{||| \cdot |||}(x) = ||| \operatorname{diag}(x) |||$$

is a symmetric gauge function. Conversely, if $\Phi : \mathbb{R}^n \to \mathbb{R}_+$ is a symmetric gauge function, then the function on $\mathbb{C}_{n \times n}$ defined by

$$||| A |||_{\Phi} = \Phi(s(A))$$

is a unitarily invariant norm.

Proof. Suppose $\|\|\cdot\|\|$ is a unitarily invariant norm on $\mathbb{C}_{n\times n}$ and let $\Phi_{\|\|\cdot\|\|}(x) = \|\|\operatorname{diag}(x)\|\|$ for all $x \in \mathbb{R}^n$. Then $\Phi_{\|\|\cdot\|\|}$ is a norm on $\mathbb{R}^n$, since $\|\|\cdot\|\|$ is a norm. For all $P \in \mathbb{S}_n$ and $x \in \mathbb{R}^n$

$$\Phi_{\|\|\cdot\|\|}(Px) = \|\|\operatorname{diag}(Px)\|\| = \|\|P\operatorname{diag}(x)\|\| = \|\|\operatorname{diag}(x)\|\|,$$

because $P \in \mathbb{S}_n \subset \mathrm{U}_n$ and $\|\|\cdot\|\|$ is unitarily invariant. If $\varepsilon_i = \pm 1$ for $1 \leqslant i \leqslant n$ and $x = (x_1, \ldots, x_n) \in \mathbb{R}^n$, then

$$\Phi_{\|\|\cdot\|\|}(\varepsilon_1 x_1, \ldots, \varepsilon_n x_n) = \|\|\operatorname{diag}(\varepsilon_1, \ldots, \varepsilon_n)\operatorname{diag}(x)\|\| = \|\|\operatorname{diag}(x)\|\|$$

because $\operatorname{diag}(\varepsilon_1, \ldots, \varepsilon_n) \in \mathrm{U}_n$.

For the converse, suppose $\Phi : \mathbb{R}^n \to \mathbb{R}_+$ is a symmetric gauge function and let $\|\|\cdot\|\|_\Phi : \mathbb{C}_{n\times n} \to \mathbb{R}$ be defined by

$$\|\|A\|\|_\Phi = \Phi(s(A)).$$

Because Φ is a norm, $\|\|A\|\|_\Phi \geqslant 0$ for all $A \in \mathbb{C}_{n\times n}$, and $\|\|A\|\|_\Phi = 0$ if and only if $s(A) = 0$ if and only if $A = 0$. If $\alpha \in \mathbb{C}$ and $A \in \mathbb{C}_{n\times n}$, then

$$\|\|\alpha A\|\|_\Phi = \Phi(s(\alpha A)) = \Phi(|\alpha|s(A)) = |\alpha|\Phi(s(A)) = |\alpha|\|\|A\|\|_\Phi.$$

For all $A, B \in \mathbb{C}_{n\times n}$, it is true that (for the Ky Fan k-norms are norms)

$$s(A+B) \prec_w s(A) + s(B).$$

It then follows from Theorem 1.9 that

$$\begin{aligned}\|\|A+B\|\|_\Phi &= \Phi(s(A+B)) \\ &\leqslant \Phi(s(A) + s(B)) \\ &\leqslant \Phi(s(A)) + \Phi(s(B)) \\ &= \|\|A\|\|_\Phi + \|\|B\|\|_\Phi\end{aligned}$$

for all $A, B \in \mathbb{C}_{n\times n}$. Thus $\|\|\cdot\|\|_\Phi$ is a norm on $\mathbb{C}_{n\times n}$. It is obviously unitarily invariant because $s(UAV) = s(A)$ for all $A \in \mathbb{C}_{n\times n}$ and all $U, V \in \mathrm{U}_n$. □

Combining Theorem 1.9 and Theorem 1.10 yields the following important result, which relates inequalities for unitarily invariant norms and weak majorization.

Theorem 1.11. (Fan Dominance Theorem) *Let $A, B \in \mathbb{C}_{n\times n}$. Then*

$$\|\|A\|\| \leqslant \|\|B\|\|$$

for all unitarily invariant norms $\|\|\cdot\|\|$ if and only if

$$s(A) \prec_w s(B),$$

that is, if and only if

$$\|A\|_{(k)} \leqslant \|B\|_{(k)}$$

for all Ky Fan k-norms $\|\cdot\|_{(k)}$, where $1 \leqslant k \leqslant n$.

Notes and References. The proofs of Theorem 1.9 and Theorem 1.10 are adopted from [Mir60] and [Bha97, p.91], respectively.

1.4 The Matrix Exponential Map

The *matrix exponential map* $\exp : \mathbb{C}_{n \times n} \to \mathbb{C}_{n \times n}$ is defined as

$$\exp(X) = \sum_{n=0}^{\infty} \frac{X^n}{n!} = I + X + \frac{X^2}{2!} + \frac{X^3}{3!} + \cdots$$

It is well-defined, because for any matrix norm $\|\cdot\|$ on $\mathbb{C}_{n \times n}$,

$$\sum_{n=0}^{\infty} \left\| \frac{X^n}{n!} \right\| \leq \sum_{n=0}^{\infty} \frac{\|X\|^n}{n!} = e^{\|X\|}.$$

We also denote $\exp(X)$ by e^X.

It follows from the spectral theorem for Hermitian matrices that

$$\exp : \mathbb{H}_n \to \mathbb{P}_n \text{ is bijective.} \tag{1.27}$$

The following theorem collects some basic properties of the matrix exponential map.

Theorem 1.12. *The following statements are true for all $X, Y \in \mathbb{C}_{n \times n}$.*

(1) $e^0 = I$.

(2) $\left(e^X\right)^* = e^{X^*}$.

(3) If $S \in \mathbb{C}_{n \times n}$ is invertible, then $e^{SXS^{-1}} = Se^X S^{-1}$.

(4) $\det\left(e^X\right) = e^{\operatorname{tr} X}$.

(5) e^X is invertible.

(6) If $XY = YX$, then $e^{X+Y} = e^X e^Y = e^Y e^X$.

(7) $(e^X)^{-1} = e^{-X}$.

(8) If $X, Y \in \mathbb{H}_n$, then $e^X e^Y = e^Y e^X$ if and only if $XY = YX$.

Proof. We only show (6) and (8), because all others are either straightforward by definition or by the preceding statement.

(6). Suppose $XY = YX$. Then

$$(X+Y)^n = \sum_{k=0}^{n} \frac{n!}{k!(n-k)!} X^k Y^{n-k}.$$

Thus

$$e^X e^Y = \left(\sum_{n=0}^{\infty} \frac{X^n}{n!}\right)\left(\sum_{n=0}^{\infty} \frac{Y^n}{n!}\right)$$
$$= \sum_{n=0}^{\infty} \sum_{k=0}^{n} \frac{X^k}{k!} \frac{Y^{n-k}}{(n-k)!}$$
$$= \sum_{n=0}^{\infty} \frac{1}{n!} \sum_{k=0}^{n} \frac{n!}{k!(n-k)!} X^k Y^{n-k}$$
$$= \sum_{n=0}^{\infty} \frac{1}{n!}(X+Y)^n$$
$$= e^{X+Y}.$$

(8). It suffices to show the necessity by (6). Suppose $X, Y \in \mathbb{H}_n$ and $e^X e^Y = e^Y e^X$. Then there exists $U \in \mathbb{U}_n$ such that both $Ue^X U^*$ and $Ue^Y U^*$ are real diagonal. It follows from (3) and (1.27) that UXU^* and UYU^* are also both real diagonal, and thus they commute. Now

$$UXYU^* = (UXU^*)(UYU^*) = (UYU^*)(UXU^*) = UYXU^*.$$

Therefore, $XY = YX$. □

The following famous result is very useful in proving matrix exponential inequalities.

Theorem 1.13. (Lie Product Formula) *For any $X, Y \in \mathbb{C}_{n \times n}$,*

$$e^{X+Y} = \lim_{m \to \infty} \left(e^{X/m} e^{Y/m}\right)^m. \tag{1.28}$$

Proof. Let $\|\cdot\|$ be any norm on $\mathbb{C}_{n \times n}$. For any $A, B \in \mathbb{C}_{n \times n}$ and for each $m \in \mathbb{N}$, we have

$$A^m - B^m = \sum_{k=0}^{m-1} A^{m-1-k}(A-B)B^k.$$

Thus if $M = \max\{\|A\|, \|B\|\}$, then

$$\|A^m - B^m\| \leqslant mM^{m-1}\|A - B\|. \tag{1.29}$$

Because $e^{(X+Y)/m} = I + \dfrac{X+Y}{m} + O\left(\dfrac{1}{m^2}\right)$ and

$$e^{X/m} e^{Y/m} = \left(I + \frac{X}{m} + O\left(\frac{1}{m^2}\right)\right)\left(I + \frac{Y}{m} + O\left(\frac{1}{m^2}\right)\right)$$
$$= I + \frac{X}{m} + \frac{Y}{m} + O\left(\frac{1}{m^2}\right),$$

we see
$$e^{(X+Y)/m} - e^{X/m}e^{Y/m} = O\left(\frac{1}{m^2}\right).$$

By (1.29) and the fact that $\max\{\|e^{(X+Y)/m}\|, \|e^{X/m}e^{Y/m}\|\} \leq e^{(\|A\|+\|B\|)/m}$, we have
$$\left\|e^{X+Y} - \left(e^{X/m}e^{Y/m}\right)^m\right\| \leq me^{\|A\|+\|B\|}O\left(\frac{1}{m^2}\right),$$
which yields the theorem. □

The following is a variation of the Lie product formula (1.28).

Theorem 1.14. *For any $X, Y \in \mathbb{C}_{n\times n}$,*
$$e^{X+Y} = \lim_{m\to\infty}\left(e^{X/2m}e^{Y/m}e^{X/2m}\right)^m. \tag{1.30}$$

Proof. For any $X, Y \in \mathbb{C}_{n\times n}$, we have
$$\begin{aligned}\lim_{m\to\infty}\left(e^{X/2m}e^{Y/m}e^{X/2m}\right)^m &= \lim_{m\to\infty}\left(e^{-X/2m}\left(e^{X/m}e^{Y/m}\right)e^{X/2m}\right)^m\\ &= \lim_{m\to\infty}e^{-X/2m}\left(e^{X/m}e^{Y/m}\right)^m e^{X/2m}\\ &= \lim_{m\to\infty}e^{-X/2m}\lim_{m\to\infty}\left(e^{X/m}e^{Y/m}\right)^m \lim_{m\to\infty}e^{X/2m}\\ &= Ie^{X+Y}I \quad \text{by (1.28)}\\ &= e^{X+Y}.\end{aligned}$$
This completes the proof. □

Notes and References. The proof of Theorem 1.13 is adopted from [Bha97, p.254]. See [How83, p.613] for another proof. A proof without using norm is available in [Hal03, p.35]. See [CFKK82, p.60] for interesting discussions about the name "Lie Product Formula".

1.5 Compound Matrices and Applications

In this section, we first review some nice properties of compound matrices, and then apply them to derive several classic matrix inequalities. The techniques used here will be applied repeatedly in later chapters.

1.5.1 Compound Matrices

Suppose $A \in \mathbb{C}_{n \times n}$. For index sets $\alpha \subset \{1,\ldots,n\}$ and $\beta \subset \{1,\ldots,n\}$, we denote by $A[\alpha|\beta]$ the submatrix of A whose entries lie in the rows of A indexed by α and the columns indexed by β. If $\alpha = \beta$, the principal submatrix $A[\alpha|\alpha]$ of A is abbreviated as $A[\alpha]$. For all $1 \leqslant k \leqslant n$, the *kth (multiplicative) compound* of A is defined as the $\binom{n}{k} \times \binom{n}{k}$ complex matrix $C_k(A)$ whose elements are given by

$$C_k(A)_{\alpha,\beta} = \det A[\alpha|\beta], \tag{1.31}$$

where $\alpha, \beta \in Q_{k,n}$ and

$$Q_{k,n} = \{\omega = (\omega(1),\ldots,\omega(k)) : 1 \leqslant \omega(1) < \cdots < \omega(k) \leqslant n\}$$

is the set of increasing sequences of length k chosen from $\{1,\ldots,n\}$, and $A[\alpha|\beta]$ is the submatrix of A whose rows and columns are indexed by α and β, respectively. In particular, $C_1(A) = A$ and $C_n(A) = \det A$.

For example, if $A = \begin{pmatrix} 1 & 2 & 3 \\ 4 & 5 & 6 \\ 7 & 8 & 9 \end{pmatrix}$, then

$$C_2(A) = \begin{pmatrix} \det A[1,2|1,2] & \det A[1,2|1,3] & \det A[1,2|2,3] \\ \det A[1,3|1,2] & \det A[1,3|1,3] & \det A[1,3|2,3] \\ \det A[2,3|1,2] & \det A[2,3|1,3] & \det A[2,3|2,3] \end{pmatrix}$$
$$= \begin{pmatrix} -3 & -6 & -3 \\ -6 & -12 & -6 \\ -3 & -6 & -3 \end{pmatrix}.$$

Compound matrices have many nice properties. The following are some related to our discussion, while a much longer list is collected in [Ber09, p. 411–412]. See [Mar73, Mar75, Mer97] for proofs.

Theorem 1.15. *Let $A, B \in \mathbb{C}_{n \times n}$. Let $s(A) = (s_1,\ldots,s_n)$ and $\lambda(A) = (\lambda_1,\ldots,\lambda_n)$ denote the vector of singular values of A in decreasing order and the vector of eigenvalues of A whose absolute values are in decreasing order, respectively. Then the following statements are true.*

(1) $C_k(A^*) = [C_k(A)]^*$.

(2) If $A = (a_{ij})$ is upper triangular, then so is $C_k(A)$ and its diagonal entries are

$$\prod_{j=1}^{k} a_{\omega(j),\omega(j)}$$

for all $\omega \in Q_{k,n}$.

(3) The eigenvalues of $C_k(A)$ are
$$\prod_{j=1}^{k} \lambda_{\omega(j)}$$
for all $\omega \in Q_{k,n}$.

(4) The singular values of $C_k(A)$ are
$$\prod_{j=1}^{k} s_{\omega(j)}$$
for all $\omega \in Q_{k,n}$.

(5) If A is unitary, then so is $C_k(A)$.

(6) If A is positive semidefinite, then so is $C_k(A)$.

(7) $C_k(AB) = C_k(A)C_k(B)$, which is called the Binet-Cauchy Theorem. Thus the map $C_k : \mathrm{GL}_n(\mathbb{C}) \to \mathrm{GL}_{\binom{n}{k}}(\mathbb{C})$ is a group homomorphism.

(8) If $A = UP$ is a polar decomposition, then so is $C_k(A) = C_k(U)C_k(P)$.

(9) If $A = QR$ is a QR decomposition, then so is $C_k(A) = C_k(Q)C_k(R)$.

1.5.2 Additive Compound Matrices

For $1 \leqslant k \leqslant n$, the kth additive compound of $A \in \mathbb{C}_{n \times n}$ is defined by

$$\Delta_k(A) = \left.\frac{d}{dt}\right|_{t=0} C_k(I + tA). \tag{1.32}$$

An explicit formula for Δ_k is given in [Sch70, Theorem 1] and [Fie74, Theorem 2.4].

Since
$$C_k(I + tA)C_k(I + tB) = C_k((I + tA)(I + tB))$$
$$= C_k(I + t(A + B) + t^2 AB),$$

we have
$$\Delta_k(A + B) = \left.\frac{d}{dt}\right|_{t=0} C_k(I + t(A + B) + t^2 AB)$$
$$= \left.\frac{d}{dt}\right|_{t=0} C_k(I + tA)C_k(I + tB)$$
$$= \left(\left.\frac{d}{dt}\right|_{t=0} C_k(I + tA)\right) I + I \left(\left.\frac{d}{dt}\right|_{t=0} C_k(I + tB)\right)$$
$$= \Delta_k(A) + \Delta_k(B).$$

Thus $\Delta_k : \mathbb{C}_{n \times n} \to \mathbb{C}_{\binom{n}{k} \times \binom{n}{k}}$ is a linear map.

There is a connection between compound matrices and ordinary differential equations. Suppose $X(t)$ is a solution to the system

$$\frac{dx}{dt} = A(t)x,$$

where $A(t) \in \mathbb{C}_{n \times n}$ is a continuous matrix valued function of t. Then $X(t+s) = (I + sA(t))X(t) + o(s)$ for $s \to 0$. By Theorem 1.15, we have for all $1 \leq k \leq n$,

$$C_k(X(t+s)) = C_k(I + sA(t))C_k(X(t)) + o(s).$$

Taking $\left.\dfrac{d}{ds}\right|_{s=0}$ on both sides yields

$$\frac{d}{dt} C_k(X(t)) = \Delta_k(A(t))C_k(X(t)).$$

In other words, $Y(t) = C_k(X(t))$ is a solution to $\dfrac{dy}{dt} = \Delta_k(A(t))y$. Therefore, we have for all $A \in \mathbb{C}_{n \times n}$,

$$e^{\Delta_k(A)} = C_k(e^A). \tag{1.33}$$

Since the group homomorphism $C_k : \mathrm{GL}_n(\mathbb{C}) \to \mathrm{GL}_{\binom{n}{k}}(\mathbb{C})$ is continuous, it is actually a Lie group homomorphism by [Hel78, p.117]. Its differential dC_k at the identity is given by

$$dC_k(A) = \left.\frac{d}{dt}\right|_{t=0} C_k(e^{tA}),$$

which is equal to $\Delta_k(A)$ by (1.33). Thus $\Delta_k : \mathbb{C}_{n \times n} \to \mathbb{C}_{\binom{n}{k} \times \binom{n}{k}}$ is the derived Lie algebra homomorphism, where the Lie bracket operation on $\mathbb{C}_{n \times n}$ is given by

$$[A, B] = AB - BA.$$

Additive compound matrices have the following nice properties.

Theorem 1.16. *The following are true for any $A = (a_{ij}) \in \mathbb{C}_{n \times n}$.*

(1) $\Delta_k(A^*) = \Delta_k(A)^*$.

(2) The diagonal entries of $\Delta_k(A)$ are

$$\sum_{j=1}^{k} a_{\omega(j), \omega(j)}$$

for all $\omega \in Q_{k,n}$.

(3) If the eigenvalues of A are $\lambda_1, \ldots, \lambda_n$, then the eigenvalues of $\Delta_k(A)$ are

$$\sum_{j=1}^{k} \lambda_{\omega(j)}$$

for all $\omega \in Q_{k,n}$. Consequently, $\operatorname{tr} \Delta_k(A) = \binom{n-1}{k-1} \operatorname{tr} A$.

1.5.3 Applications to Matrix Inequalities

Let $A \in \mathbb{C}_{n \times n}$. There are two real n-vectors naturally associated with A, namely, $s(A)$ consisting of singular values of A and $|\lambda(A)|$ consisting of the absolute values of eigenvalues of A, both in decreasing order. If A is Hermitian, then there are two additional real n-vectors: $d(A)$ consisting of diagonal entries and $\lambda(A)$. If A is invertible, let $A = QR$ be the unique QR decomposition and let $A = L\omega U$ be a Gelfand-Naimark decomposition. Then there are two additional real n-vectors: $r(A) := \operatorname{diag} R$ and $|u(A)| := |\operatorname{diag} U|$ as in (1.17). What is interesting is that the relationships among these vectors can be characterized in terms of majorizations. Moreover, some of these majorizations can be generalized to Lie groups.

In this section, we apply Theorem 1.15 and Theorem 1.16 to prove the following results:

(1) the Weyl-Horn inequality on $s(A)$ and $|\lambda(A)|$,

(2) the Kostant inequality on $s(A)$ and $r(A)$,

(3) the Huang-Tam inequality on $r(A)$ and $|u(A)|$,

(4) the Yamamoto theorem on $s_i(A)$ and $\lambda_i(A)$ for all $1 \leqslant i \leqslant n$, and

(5) the Schur-Horn inequality on $d(A)$ and $\lambda(A)$.

The following result is given by Weyl [Wey49].

Theorem 1.17. (Weyl) *If $A \in \mathbb{C}_{n \times n}$, then $|\lambda(A)| \prec_{\log} s(A)$.*

Proof. Let $\lambda(A) = (\lambda_1, \ldots, \lambda_n)$ with $|\lambda_1| \geqslant \cdots \geqslant |\lambda_n|$ and let $s(A) = (s_1, \ldots, s_n)$ with $s_1 \geqslant \cdots \geqslant s_n$. If $x \in \mathbb{C}^n$ is a unit eigenvector of A associated with λ_1, then

$$|\lambda_1| = \|\lambda_1 x\|_2 = \|Ax\|_2 \leqslant \max_{\|x\|_2 = 1} \|Ax\|_2 = s_1.$$

By Theorem 1.15, for each $1 \leqslant k \leqslant n$

$$\prod_{j=1}^{k} |\lambda_j| = |\lambda_1|(C_k(A)) \leqslant s_1(C_k(A)) = \prod_{j=1}^{k} s_j.$$

If $A = U\text{diag}(s_1, \ldots, s_n)V$ is a singular value decomposition, then
$$\prod_{j=1}^n |\lambda_j| = |\det A| = |\det U| \cdot \prod_{j=1}^n s_j \cdot |\det V| = \prod_{j=1}^n s_j.$$
Therefore, $|\lambda(A)| \prec_{\log} s(A)$. □

See Theorem 3.21 for an extension of Theorem 1.17 in Lie groups. The converse of Theorem 1.17 is also true and is due to A. Horn [Hor54b].

Theorem 1.18. (A. Horn) *Let $\lambda = (\lambda_1, \ldots, \lambda_n) \in \mathbb{C}^n$ and $s = (s_1, \ldots, s_n) \in \mathbb{R}_+^n$. If $|\lambda| \prec_{\log} s$, then there exists $A \in \mathbb{C}_{n \times n}$ such that $\lambda_1, \ldots, \lambda_n$ are eigenvalues of A and $s_1, \ldots, s_n$ are singular values of A.*

Proof. Without loss of generality, we assume that $|\lambda_1| \geq \cdots \geq |\lambda_n|$ and $s_1 \geq \cdots \geq s_n \geq 0$. We divide the proof into two cases: A is nilpotent or not.

Case 1: $\lambda_1 = 0$. Then $\lambda = 0$ and $s_n = 0$. So we choose
$$A = \begin{pmatrix} 0 & s_1 & & \\ & \ddots & \ddots & \\ & & \ddots & s_{n-1} \\ & & & 0 \end{pmatrix}.$$

Case 2: $\lambda_1 \neq 0$. We will use induction on n. When $n = 2$, the matrix
$$A = \begin{pmatrix} \lambda_1 & \mu \\ 0 & \lambda_2 \end{pmatrix}$$
has singular values s_1 and s_2 if we set
$$\mu = (s_1^2 + s_2^2 - |\lambda_1|^2 - |\lambda_2|^2)^{1/2}.$$
Suppose that the statement is true for $\lambda_1 \neq 0$ when $n = m \geq 2$. Let $n = m+1$ and let $j \geq 2$ be the largest index such that $s_{j-1} \geq |\lambda_1| \geq s_j$. Clearly
$$s_1 \geq \max\{|\lambda_1|, s_1 s_j/|\lambda_1|\} \geq \min\{|\lambda_1|, s_1 s_j/|\lambda_1|\}.$$
Then there exist 2×2 unitary matrices U_1 and V_1 such that
$$U_1 \begin{pmatrix} s_1 & \\ & s_j \end{pmatrix} V_1 = \begin{pmatrix} \lambda_1 & \mu' \\ 0 & s_1 s_j/|\lambda_1| \end{pmatrix},$$
where $\mu' = (s_1^2 + s_j^2 - |\lambda_1|^2 - s_1^2 s_j^2/|\lambda_1|^2)^{1/2}$. Set
$$U_2 = U_1 \oplus I_{m-1},$$
$$V_2 = V_1 \oplus I_{m-1},$$
$$A_1 = U_2 \, \text{diag}(s_1, s_j, s_2, \ldots, s_{j-1}, s_{j+1}, \ldots, s_{m+1}) V_2$$
$$= \begin{pmatrix} \lambda_1 & \mu' \\ 0 & s_1 s_j/|\lambda_1| \end{pmatrix} \oplus \text{diag}(s_2, \ldots, s_{j-1}, s_{j+1}, \ldots, s_{m+1}).$$

Note that

$$\prod_{i=2}^{k} |\lambda_i| \leqslant |\lambda_1|^{k-1} \leqslant \prod_{i=2}^{k} s_i, \qquad \forall\, k = 2, \ldots, j-1,$$

$$\prod_{i=2}^{k} |\lambda_i| = \frac{1}{|\lambda_1|} \prod_{i=1}^{k} |\lambda_i| \leqslant \frac{s_1 s_j}{|\lambda_1|} \prod_{i=2, i\neq j}^{k} s_i, \qquad \forall\, k = j, \ldots, m,$$

$$\prod_{i=2}^{m+1} |\lambda_i| = \frac{s_1 s_j}{|\lambda_1|} \prod_{i=2, i\neq j}^{m+1} s_i.$$

This means that

$$|(\lambda_2, \ldots, \lambda_{m+1})| \prec_{\log} (s_1 s_j/|\lambda_1|, s_2, \ldots, s_{j-1}, s_{j+1}, \ldots, s_{m+1}).$$

If $\lambda_2 = 0$, we apply Case 1; if $\lambda_2 \neq 0$, we apply the induction hypothesis. In any case, there exist $m \times m$ unitary matrices U_3 and V_3 such that

$$U_3 \,\mathrm{diag}\left(\frac{s_1 s_2}{|\lambda_1|}, s_2, \ldots, s_{j-1}, s_{j+1}, \ldots, s_{m+1}\right) V_3$$

is upper triangular with diagonal $(\lambda_2, \ldots, \lambda_{m+1})$. Then

$$A = U_4 A_1 V_4$$

is the desired matrix, where $U_4 = 1 \oplus U_3$ and $V_4 = 1 \oplus V_3$. $\square$

The following result is given by Kostant [Kos73] in the context of Lie groups.

Theorem 1.19. (Kostant) *Let $A \in \mathrm{GL}_n(\mathbb{C})$ and let $A = QR$ be the unique QR decomposition with $r(A) := \mathrm{diag}\, R > 0$. Then $r(A) \prec_{\log} s(A)$.*

Proof. For each $1 \leqslant i \leqslant n$, let A_i denote the i-th column of A and let e_i denote the unit vector with 1 on the i-th entry and 0 elsewhere. Denote $r(A) = (r_1, \ldots, r_n)$. Note that

$$r_i \leqslant \|QR_i\|_2 = \|A_i\|_2 = \|Ae_i\|_2 \leqslant \max_{\|x\|_2 = 1} \|Ax\|_2 = s_1$$

for all $1 \leqslant i \leqslant n$. Thus

$$\max_{1 \leqslant i \leqslant n} r_i \leqslant s_1.$$

By Theorem 1.15, $C_k(A) = C_k(Q) C_k(R)$ is the QR decomposition of $C_k(A)$. Thus for $1 \leqslant k \leqslant n$,

$$\max_{1 \leqslant i_1 < \cdots < i_k \leqslant n} \prod_{j=1}^{k} r_{i_j} = \max_{\omega \in Q_{k,n}} r_\omega(C_k(A)) \leqslant s_1(C_k(A)) = \prod_{j=1}^{k} s_j$$

and
$$\prod_{j=1}^{n} r_j = \det R = |\det Q| \cdot \det R = |\det A| = \prod_{j=1}^{n} s_j.$$

Therefore, $r(A) \prec_{\log} s(A)$. $\square$

There is another (easier) way to prove Theorem 1.19:
$$r(A) = \lambda(R) \prec s(R) = s(Q^{-1}A) = s(A),$$
as an application of Theorem 1.17. The converse of Theorem 1.19 is also true.

Theorem 1.20. (Kostant) *Let $r = (r_1, \ldots, r_n) > 0$ and $s = (s_1, \ldots, s_n) > 0$. If $r \prec_{\log} s$, then there exists $A \in \mathrm{GL}_n(\mathbb{C})$ such that $r(A) = r = \mathrm{diag}\, R$, where $A = QR$ is the unique QR decomposition of A, and $s(A) = s$ is the vector of singular values of A.*

Proof. Suppose $r \prec_{\log} s$. By Theorem 1.18, there exists $A' \in \mathrm{GL}_n(\mathbb{C})$ such that
$$\lambda(A') = r \quad \text{and} \quad s(A) = s.$$

By Schur's triangularization theorem, there is a unitary matrix $Q \in \mathrm{U}(n)$ such that
$$Q^* A' Q = \begin{pmatrix} r_1 & * & * \\ & \ddots & * \\ & & r_n \end{pmatrix} =: R.$$

So $A := A'U = QR$ is the desired matrix with $r(A) = r$ and $s(A) = s$. $\square$

For $a = (a_1, \ldots, a_n) \in \mathbb{R}^n$ and $b = (b_1, \ldots, b_n) \in \mathbb{R}^n$, we write $a \triangleleft b$ if
$$\sum_{i=1}^{k} a_i \leq \sum_{i=1}^{k} b_i, \quad \forall k = 1, \ldots, n-1,$$
$$\sum_{i=1}^{n} a_i = \sum_{i=1}^{n} b_i.$$

Similarly, for $a, b \in \mathbb{R}_+^n$, we write $a \triangleleft_{\log} b$ if
$$\prod_{i=1}^{k} a_i \leq \prod_{i=1}^{k} b_i, \quad \forall k = 1, \ldots, n-1,$$
$$\prod_{i=1}^{n} a_i = \prod_{i=1}^{n} b_i.$$

Note that there is no rearrangement of entries of vectors in the definition of $\triangleleft$ or $\triangleleft_{\log}$. Thus the partial order $\triangleleft$ is different from majorization $\prec$. However, if both a and b are in decreasing order, then $a \triangleleft b$ if and only if $a \prec b$.

Review of Matrix Theory

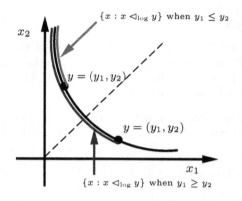

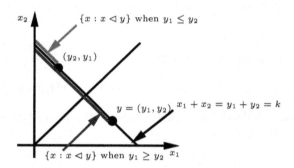

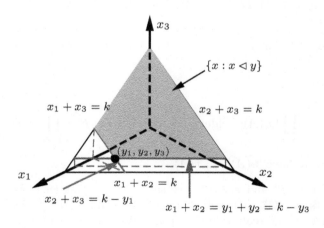

The first diagram gives the region for $\triangleleft_{\log}$ when $n = 2$. The second and third diagrams give the regions for $\triangleleft$ when $n = 2$ and $n = 3$, restricted to $\mathbb{R}_+^2$ and $\mathbb{R}_+^3$, respectively.

Let $\mathbb{F} = \mathbb{C}$ or $\mathbb{R}$. Given $A \in \mathrm{GL}_n(\mathbb{F})$, let $r(A) = \mathrm{diag}\, R$, where $A = QR$ is the unique QR decomposition of A such that the diagonal entries of R are positive, and let $u(A) = \mathrm{diag}\, U$, where $A = L\omega U$ is a Gelfand-Naimark decomposition of A as in Theorem 1.4.

The following two theorems, both by Huang and Tam [HT10], give a complete relation between $r(A)$ and $u(A)$ for all $A \in \mathrm{GL}_n(\mathbb{F})$, in terms of the partial order $\triangleleft_{\log}$.

Theorem 1.21. (Huang-Tam) *If $A \in \mathrm{GL}_n(\mathbb{F})$, then $|u(A)| \triangleleft_{\log} r(A)$.*

Proof. Suppose $A = QR$ and $A = L\omega U$ are the QR and Gelfand-Naimark decompositions of $A \in \mathrm{GL}_n(\mathbb{F})$, respectively. Write $r(A) = (r_1(A), \ldots, r_n(A))$ and $u(A) = (u_1(A), \ldots, u_n(A))$. Since $u_1(A)$ is the first nonzero entry of the first column of A and $r_1(A)$ is the Euclidean norm of the first column of A, we have

$$|u_1(A)| \leqslant r_1(A). \tag{1.34}$$

It is easy to see from Theorem 1.15 that

$$C_k(A) = C_k(Q)C_k(R) \quad \text{and} \quad C_k(A) = C_k(L)C_k(\omega)C_k(U)$$

are the QR and Gelfand-Naimark decompositions of $C_k(A)$, respectively. Note also that for all $1 \leqslant k \leqslant n$

$$u_1(C_k(A)) = \prod_{i=1}^{k} u_i(A) \quad \text{and} \quad r_1(C_k(A)) = \prod_{i=1}^{k} r_i(A).$$

An application of (1.34) to $C_k(A)$ yields that

$$\prod_{i=1}^{k} |u_i(A)| = |u_1(C_k(A))| \leqslant r_1(C_k(A)) = \prod_{i=1}^{k} r_i(A), \qquad \forall\, 1 \leqslant k \leqslant n.$$

Moreover,

$$\prod_{i=1}^{n} |u_i(A)| = |\det U| = |\det A| = \det R = \prod_{i=1}^{n} r_i(A).$$

Therefore, we have $|u(A)| \triangleleft_{\log} r(A)$. $\square$

Given $u \in \mathbb{F}^n$ and $r \in \mathbb{R}_+^n$, we say that the pair (u, r) is $\mathbb{F}$-*realizable* if there exists $A \in \mathrm{GL}_n(\mathbb{F})$ such that $u(A) = u$ and $r(A) = r$. It is not hard to see that (u, r) is $\mathbb{F}$-realizable if and only if there exists $Q \in \mathrm{U}_n(\mathbb{F})$ such that $u = u(Q \, \mathrm{diag}\, r)$. We remark that

(1) if (u, r) is $\mathbb{C}$-realizable, then so is $(D_\theta u, r)$ for all $\theta = (\theta_1, \ldots, \theta_n) \in \mathbb{R}^n$, where $D_\theta = \text{diag}(e^{i\theta_1}, \ldots, e^{i\theta_n})$.

(2) if (u, r) is $\mathbb{R}$-realizable, then so is (Du, r), where $D = \text{diag}(\pm 1, \ldots, \pm 1)$.

The reason for (1) is that if $Q \, \text{diag} \, r = L\omega U$, by (1.16) we get

$$\text{diag}(e^{i\theta_{\omega^{-1}(1)}}, \ldots, e^{i\theta_{\omega^{-1}(n)}}) Q \, \text{diag} \, a = \text{diag}(e^{i\theta_{\omega^{-1}(1)}}, \ldots, e^{i\theta_{\omega^{-1}(n)}}) L\omega U$$
$$= L' \text{diag}(e^{i\theta_{\omega^{-1}(1)}}, \ldots, e^{i\theta_{\omega^{-1}(n)}}) \omega U$$
$$= L' (\omega \text{diag}(e^{i\theta_1}, \ldots, e^{i\theta_n}) \omega^{-1}) \omega U$$
$$= L' \omega \text{diag}(e^{i\theta_1}, \ldots, e^{i\theta_n}) U,$$

where $L' = \text{diag}(e^{i\theta_{\omega^{-1}(1)}}, \ldots, e^{i\theta_{\omega^{-1}(n)}}) L \, \text{diag}(e^{-i\theta_{\omega^{-1}(1)}}, \ldots, e^{-i\theta_{\omega^{-1}(n)}})$ is still unit lower triangular. The real case (2) is similar.

Theorem 1.22. (Huang-Tam) *Let* $r = (r_1, \ldots, r_n) > 0$ *and* $u = (u_1, \ldots, u_n) \in \mathbb{F}^n$. *If* $|u| \triangleleft_{\log} r$, *then there exists* $A \in \text{GL}_n(\mathbb{F})$ *such that* $r(A) = r$ *and* $u(A) = u$ *and* A *has an LU decomposition. Indeed, if* $u > 0$ *and* $u \triangleleft_{\log} r$, *then there exists* $Q \in \text{SO}(n)$ *such that* $u((Q \, \text{diag} \, r)) = u$ *and* $Q \, \text{diag} \, r$ *has an LU decomposition.*

Proof. We proceed by induction on n. Because of the above remark it is sufficient to consider $u > 0$ and prove the last statement.

When $n = 2$, suppose $(u_1, u_2) \triangleleft_{\log} (r_1, r_2)$, that is, $u_1 \leqslant r_1$ and $u_1 u_2 = r_1 r_2$. Let

$$Q = \begin{pmatrix} p & -\sqrt{1-p^2} \\ \sqrt{1-p^2} & p \end{pmatrix} \in \text{SO}(2),$$

where $p = u_1/r_1 \in (0, 1]$. The first column of

$$A = Q \, \text{diag}(r_1, r_2) = \begin{pmatrix} r_1 p & -r_2 \sqrt{1-p^2} \\ r_1 \sqrt{1-p^2} & r_2 p \end{pmatrix}$$

has $r_1 p = u_1$ (which is nonzero) as its first entry, so A has an LU decomposition. Clearly $r(A) = r$. Since $u_1 u_2 = r_1 r_2$, we have $u(A) = u$ by Theorem 1.21.

Suppose that the statement is true for $n \leqslant k$. Now consider $n = k + 1$. Then

$$u' := (u_1, \ldots, u_{k-1}, \frac{r_1 \cdots r_k}{u_1 \cdots u_{k-1}}) \triangleleft_{\log} (r_1, \ldots, r_k) =: r'.$$

By the induction hypothesis there exists $Q' \in \text{SO}(k)$ such that

$$A' = Q' \, \text{diag}(r_1, \ldots, r_k)$$

has an LU decomposition satisfying $u(A') = u'$ and $r(A') = r'$. Set

$$Q_2 := \begin{pmatrix} t & -\sqrt{1-t^2} \\ \sqrt{1-t^2} & t \end{pmatrix} \in \text{SO}(2),$$

where $t = \dfrac{u_1 \cdots u_k}{r_1 \cdots r_k} \leqslant 1$. The first $k-1$ rows of

$$A = (I_{k-1} \oplus Q_2)(Q' \oplus 1)\operatorname{diag} r$$

are those of $(Q' \oplus 1)\operatorname{diag} r$, so that the first $k-1$ rows of the matrices $A(k|k)$ and A' are identical and the last row of $A(k|k)$ is t times the last row of A'. So

$$\det A(k|k) = t r_1 \cdots r_k = u_1 \cdots u_k.$$

Moreover $A(k|k)$ has an LU decomposition and $u_i(A) = u_i$ for all $i = 1, \ldots, k$. Hence $A = Q \operatorname{diag} r$ has an LU decomposition and is the required matrix, where $Q = (I_{k-1} \oplus Q_2)(Q' \oplus 1) \in \mathrm{SO}(k+1)$. $\square$

It is known as Gelfand formula in [HJ13, p.349] that

$$\lim_{m \to \infty} \|A^m\|^{1/m} = \rho(A) \tag{1.35}$$

for any matrix norm $\|\cdot\|$ on $\mathbb{C}_{n \times n}$, where $\rho(A)$ denotes the spectral radius of $A \in \mathbb{C}_{n \times n}$. It is further shown in [Yam67, p.174] that if $\{m_k\}_{k \in \mathbb{N}}$ is a strictly increasing sequence of positive integers such that m_{k-1} divides m_k, then the sequence $\{\|A^{m_k}\|^{1/m_k}\}_{k \in \mathbb{N}}$ is monotonically decreasing and converges to $\rho(A)$.

Since $\|A\| = s_1(A)$ for the spectral norm and $\rho(A) = |\lambda_1(A)|$, it follows from (1.35) that

$$\lim_{m \to \infty} [s_1(A^m)]^{1/m} = |\lambda_1(A)|. \tag{1.36}$$

The following result of Yamamoto [Yam67, p.175] is a direct generalization of (1.36).

Theorem 1.23. (Yamamoto) *Let* $A \in \mathbb{C}_{n \times n}$. *Then*

$$\lim_{m \to \infty} [s_k(A^m)]^{1/m} = |\lambda_k(A)|, \quad \forall\, 1 \leqslant k \leqslant n. \tag{1.37}$$

Proof. We may assume that A is nonsingular, since the singular case can be handled by continuity argument. Then $s_1(A^m) \geqslant \cdots \geqslant s_n(A^m) > 0$ for all $m \in \mathbb{N}$ and $|\lambda_1(A)| \geqslant \cdots \geqslant |\lambda_n(A)| > 0$. Applying (1.36) on the compound matrix $C_k(A^m)$, we have for $1 \leqslant k \leqslant n$

$$\begin{aligned}
\lim_{m \to \infty} [s_1(A^m)]^{1/m} \cdots [s_k(A^m)]^{1/m} &= \lim_{m \to \infty} [s_1(A^m) \cdots s_k(A^m)]^{1/m} \\
&= \lim_{m \to \infty} [s_1(C_k(A^m))]^{1/m} \\
&= \lim_{m \to \infty} [s_1([C_k(A)]^m)]^{1/m} \\
&= |\lambda_1(C_k(A))| \\
&= \prod_{i=1}^{k} |\lambda_i(A)|.
\end{aligned}$$

Thus

$$\lim_{m\to\infty}[s_k(A^m)]^{1/m} = \frac{\lim_{m\to\infty}[s_1(A^m)]^{1/m}\cdots[s_k(A^m)]^{1/m}}{\lim_{m\to\infty}[s_1(A^m)]^{1/m}\cdots[s_{k-1}(A^m)]^{1/m}}$$
$$= \frac{\prod_{i=1}^{k}|\lambda_i(A)|}{\prod_{i=1}^{k-1}|\lambda_i(A)|}$$
$$= |\lambda_k(A)|.$$

This completes the proof. □

The following result is given by Schur [Sch23].

Theorem 1.24. (Schur) *Let $A \in \mathbb{H}_n$ and let $d(A) = (d_1, \ldots, d_n)$ denote the vector of diagonal entries of A. Then $d(A) \prec \lambda(A)$.*

Proof. By the spectral theorem for Hermitian matrices, there exists a unitary matrix $U = (u_{ij}) \in \mathbb{C}_{n \times n}$ such that $A = U^*\text{diag}(\lambda_1, \ldots, \lambda_n)U$. Then for all $1 \leqslant j \leqslant n$,

$$d_j = \sum_{i=1}^{n} \lambda_i |u_{ij}|^2 \leqslant \sum_{i=1}^{n} \lambda_1 |u_{ij}|^2 = \lambda_1.$$

So $\max_{1 \leqslant j \leqslant n} d_j \leqslant \lambda_1$. By Theorem 1.16, we have for all $1 \leqslant j \leqslant n$

$$\max_{1 \leqslant j_1 < \cdots < j_k \leqslant n} \sum_{i=1}^{k} d_{j_i} = \max_{\omega \in Q_{k,n}} d_\omega(\Delta_k(A)) \leqslant \lambda_1(\Delta_k(A)) = \sum_{j=1}^{k} \lambda_j.$$

Obviously, $\sum_{j=1}^{n} d_j = \text{tr } A = \sum_{j=1}^{n} \lambda_j$. Therefore, $d(A) \prec \lambda(A)$. □

Let $A \in \mathbb{C}_{n \times n}$. Let

$$e_k(A) = (\det A[\alpha])_{\alpha \in Q_{k,n}} = (\det A[\{1, \ldots, k\}], \ldots, \det A[\{n-k+1, \ldots, n\}])$$

denote the $\binom{n}{k}$-vector of the $k \times k$ principal minors of A arranged in lexicographical order and let

$$s_k(A) = \left(\prod_{j=1}^{k} \lambda_{\alpha(j)}\right)_{\alpha \in Q_{k,n}} = \left(\prod_{j=1}^{k} \lambda_j, \ldots, \prod_{j=n-k+1}^{n} \lambda_j\right)$$

denote the $\binom{n}{k}$-vector of the kth elementary functions of the eigenvalues of A arranged in lexicographical order. It is known [HJ13, p.54] that the sum of components of $e_k(A)$ is equal to that of $s_k(A)$ for all $1 \leqslant k \leqslant n$, being the coefficients (up to ± 1) of the characteristic polynomial of A. According to Theorem 1.15, $e_k(A)$ and $s_k(A)$ are the vector of diagonal entries of $C_k(A)$

and the vector of eigenvalues of $C_k(A)$, respectively. Therefore, Theorem 1.24 yields that
$$e_k(A) \prec s_k(A), \qquad \forall 1 \leqslant k \leqslant n \tag{1.38}$$
for all $A \in \mathbb{H}_n$.

The converse of Theorem 1.24 is true and is due to A. Horn [Hor54a].

Theorem 1.25. (A. Horn) *Let $d = (d_1, \ldots, d_n) \in \mathbb{R}^n$ and $\lambda = (\lambda_1, \ldots, \lambda_n) \in \mathbb{R}^n$. If $d \prec \lambda$, then there exists $A \in \mathbb{H}_n$ such that $\lambda(A) = \lambda$ and $d(A) = d$, where $d(A)$ denotes the vector of diagonal entries of A.*

Proof. We proceed by induction. Without loss of generality, we assume that d and λ are in decreasing order. If $n = 2$, then $d \prec \lambda$ implies $\lambda_1 \geqslant d_1 \geqslant d_2 \geqslant \lambda_2$. For the nontrivial case when $\lambda_1 > \lambda_2$,
$$U = \frac{1}{\sqrt{\lambda_1 - \lambda_2}} \begin{pmatrix} \sqrt{d_1 - \lambda_2} & -\sqrt{\lambda_1 - d_1} \\ \sqrt{\lambda_1 - d_1} & \sqrt{d_1 - \lambda_2} \end{pmatrix}$$
is a real orthogonal matrix, and the diagonal vector of $U^*(\operatorname{diag} \lambda)U$ is d.

Now assume the statement is true for all such vectors d and λ with at most $n - 1$ elements. Let $d \prec \lambda$ for $d, \lambda \in \mathbb{R}^n$. By taking k to be the smallest integer j with $1 \leqslant j < n$ such that $d_{j+1} \geqslant \lambda_{j+1}$, we have $\lambda_k \geqslant d_k \geqslant d_{k+1} \geqslant \lambda_{k+1}$. Putting
$$\lambda'_{k+1} = \lambda_k + \lambda_{k+1} - d_k,$$
we have $(d_k, \lambda'_{k+1}) \prec (\lambda_k, \lambda_{k+1})$. In the case when $n = 2$, there exists $U \in O_2$ such that $U^* \operatorname{diag}(\lambda_k, \lambda_{k+1}) U = \operatorname{diag}(d_k, \lambda'_{k+1})$. By replacing the 2×2 diagonal block consisting of (k, k), $(k, k+1)$, $(k+1, k)$, and $(k+1, k+1)$ entries of the $n \times n$ identity matrix I_n by U, we get $V \in O_n$ with
$$\operatorname{diag}(V^*(\operatorname{diag} \lambda)V) = (\lambda_1, \ldots, \lambda_{k-1}, d_k, \lambda'_{k+1}, \lambda_{k+2}, \ldots, \lambda_n) =: \lambda'.$$
That is, λ' is obtained from λ by replacing λ_k and λ_{k+1} by d_k and λ'_{k+1}, respectively. Note that $d \prec \lambda'$. Now let
$$d'' = (d_1, \ldots, d_{k-1}, d_{k+1}, \ldots, d_n) \in \mathbb{R}^{n-1},$$
$$\lambda'' = (\lambda_1, \ldots, \lambda_{k-1}, \lambda'_{k+1}, \lambda_{k+2}, \ldots, \lambda_n) \in \mathbb{R}^{n-1}.$$
Then $d'' \prec \lambda''$. Thus, by the induction hypothesis, there exists $W \in O_{n-1}$ such that
$$\operatorname{diag}(W^*(\operatorname{diag} \lambda'')W) = d''.$$
Expand W to be in O_n by inserting 1 at the (k, k) entry and 0's at all other entries of the kth row and kth column so that
$$\operatorname{diag}(W^*(\operatorname{diag} \lambda')W) = d.$$
It follows that
$$\operatorname{diag}((VW)^*(\operatorname{diag} \lambda)(VW)) = d,$$
with $VW \in O_n$. Thus, $A = (VW)^*(\operatorname{diag} \lambda)(VW)$ is the desired matrix. □

By combining Theorem 1.24 and Theorem 1.25 and making use of the Spectral Theorem for Hermitian matrices, we have the following theorem.

Theorem 1.26. (Schur-Horn) *Let* $\lambda = (\lambda_1, \ldots, \lambda_n) \in \mathbb{R}^n$ *and* $\Lambda = \operatorname{diag}(\lambda_1, \ldots, \lambda_n)$. *Then*

$$\{\operatorname{diag}(U^*\Lambda U) : U \in \mathrm{U}(n)\} = \{d \in \mathbb{R}^n : d \prec \lambda\}.$$

The above statement has an orbital interpretation: $\mathrm{U}(n)$ acts on Λ via unitary similarity and the projection of the orbit onto the diagonal is completely described by majorization.

Notes and References. The proof of Theorem 1.18 is from [Tam10a]. We note that (1.36) remains true for Hilbert space operators [Hal82, p.48]. Also see [JN90, JN93, NR90] for some generalizations of Yamamoto's theorem. The proof of Theorem 1.25 is adopted from [CL83].

Chapter 2

Structure Theory of Semisimple Lie Groups

2.1   Smooth Manifolds 41
2.2   Lie Groups and Their Lie Algebras 44
2.3   Complex Semisimple Lie Algebras 48
2.4   Real Forms 49
2.5   Cartan Decompositions 51
2.6   Root Space Decomposition 55
2.7   Iwasawa Decompositions 57
2.8   Weyl Groups 59
2.9 KA_+K Decomposition 60
2.10 Complete Multiplicative Jordan Decomposition 61
2.11 Kostant's Preorder 65

In this chapter, we summarize some analytic and algebraic structures of semisimple Lie groups and Lie algebras. Since these results are classical, we introduce them as necessary background for later chapters, providing no proofs but references. The major references are [Hum72], [Hel78], and [Kna02].

2.1   Smooth Manifolds

We begin with smooth manifolds, because a Lie group is simultaneously a smooth manifold and a group (in the algebraic sense) such that analytic structure and algebraic structure are compatible (i.e., the group operations are smooth). See [Lee13] or [War83] for a systematic introduction to smooth manifolds.

A *topological manifold* of dimension n is a second countable Hausdorff topological space of which every point has an open neighborhood that is homeomorphic to an open subset of $\mathbb{R}^n$. The following result collects some connectedness and compactness properties of topological manifolds (see [Lee13, p.7-10] for proofs).

Theorem 2.1. *Let M be a topological manifold. Then the following statements are true:*

(1) *M is locally path-connected.*

(2) *M is connected if and only if it is path-connected.*

(3) *M has countably many components.*

(4) *M is locally compact.*

(5) *M is paracompact.*

Recall that a map $F : U \to V$, where U and V are open subsets of $\mathbb{R}^n$ and $\mathbb{R}^m$, respectively, is said to be *smooth* (or C^∞) if each of the component functions of F has continuous partial derivatives of all orders.

Let M be a topological manifold of dimension n. A *coordinate chart* on M is a pair (U, φ), where $U \subset M$ is open and φ is a homeomorphism of U onto an open subset of $\mathbb{R}^n$. A *smooth structure* on M is a collection of coordinate charts $\{(U_\alpha, \varphi_\alpha) : \alpha \in \Lambda\}$ such that

(1) $\bigcup_{\alpha \in \Lambda} U_\alpha = M$,

(2) $\varphi_\alpha \circ \varphi_\beta^{-1}$ is C^∞ for all $\alpha, \beta \in \Lambda$, and

(3) the collection is maximal with respect to (2).

A topological manifold with a smooth structure is called a *smooth manifold*, or simply *manifold* unless otherwise specified. A coordinate chart on a manifold is said to be *smooth* if it is an element of the smooth structure.

For the remainder of this section, let M and N be manifolds.

If $\dim M = m$ and $\dim N = n$, then the product $M \times N$ becomes a manifold of dimension $m + n$, with a natural smooth structure that is called the product manifold structure.

A continuous map $F : M \to N$ is said to be *smooth* if for every $p \in M$, there exist smooth charts (U, φ) containing p and (V, ϕ) containing $F(p)$ such that $F(U) \subset V$ and the composite map

$$\phi \circ F \circ \varphi^{-1} : \varphi(U) \to \phi(V)$$

is C^∞. In the case that $N = \mathbb{R}$, F is called a *smooth function* on M if for every $p \in M$, there exists a smooth chart (U, φ) containing p such that $F \circ \varphi^{-1}$ is C^∞.

A smooth map $F : M \to N$ is call a *diffeomorphism* from M onto N if F is bijective and F^{-1} is smooth.

Structure Theory of Semisimple Lie Groups 43

Let $C^\infty(M)$ denote the set of all smooth functions on M. A linear map

$$v : C^\infty(M) \to \mathbb{R}$$

is called a *derivation* at $p \in M$ if it satisfies

$$v(fg) = f(p)v(g) + g(p)v(f), \qquad \forall\, f, g \in C^\infty(M).$$

The set $T_p(M)$ of all derivations of $C^\infty(M)$ at p forms a vector space, called the *tangent space* to M at p. Elements of $T_p(M)$ are called *tangent vectors* at p. For each $p \in M$, we have $\dim T_p(M) = \dim M$.

Let $F : M \to N$ be a smooth map and let $p \in M$. The *differential* of F at p is the linear map $dF_p : T_p(M) \to T_{F(p)}(N)$ defined by

$$dF_p(v)(f) = v(f \circ F), \qquad \forall\, v \in T_p(M),\, \forall\, f \in C^\infty(N).$$

The *rank* of F at $p \in M$ is the rank of dF_p. If $\operatorname{rank} F = \dim M$ at every $p \in M$, then F is called an *immersion*.

A *submanifold* of M is a subset $S \subset M$ endowed with a manifold topology and a smooth structure (i.e., S is a smooth manifold in its own right) such that the inclusion map $\iota : S \to M$ is an immersion.

The *tangent bundle* $T(M)$ of M is the disjoint union of the tangent spaces at all points of M, i.e.,

$$T(M) = \bigcup_{p \in M} T_p(M).$$

The projection map $\pi : T(M) \to M$ is defined by sending each vector in $T_p(M)$ to $p \in M$. If $\dim M = n$, then the tangent bundle $T(M)$ has a natural topology and smooth structure that make it into a $2n$-dimensional smooth manifold such that $\pi : T(M) \to M$ is a smooth map.

A *vector field* on M is a continuous map $X : M \to T(M)$ such that $X_p := X(p) \in T_p(M)$ for all $p \in M$. The set of smooth vector fields on M forms in the obvious way a vector space over $\mathbb{R}$; it is also a module over the ring $C^\infty(M)$: if X is a vector field on M and $f \in C^\infty(M)$, then $Xf \in C^\infty(M)$ is defined by $Xf(p) = X_p(f)$. Note that a vector field X on M is $\mathbb{R}$-linear on $C^\infty(M)$ and satisfies

$$X(f \cdot g) = (Xf) \cdot g + f \cdot Xg, \qquad \forall f, g \in C^\infty(M).$$

In other words, X acts as a derivation of the $\mathbb{R}$-algebra $C^\infty(M)$. In fact, derivations of $C^\infty(M)$ can be identified with smooth vector fields: A function

$$\mathcal{X} : C^\infty(M) \to C^\infty(M)$$

is a derivation if and only if it is of the form $\mathcal{X}(f) = Xf$ for some smooth vector field X on M [Lee13, p.181]. If X and Y are smooth vector fields on

M, then their composition $X \circ Y : C^\infty(M) \to C^\infty(M)$ need not be a smooth vector field in general, but the Lie bracket

$$[X, Y] := X \circ Y - Y \circ X$$

always is. The space of smooth vector fields on a manifold has the structure of a Lie algebra over $\mathbb{R}$.

2.2 Lie Groups and Their Lie Algebras

A vector space $\mathfrak{g}$ over a field $\mathbb{F}$ with a product $\mathfrak{g} \times \mathfrak{g} \to \mathfrak{g}$, denoted by $(X, Y) \mapsto [X, Y]$ that is called the *Lie bracket* of X and Y, is called a *Lie algebra* over $\mathbb{F}$ if the following three conditions are satisfied:

(1) The Lie bracket is bilinear.

(2) $[X, X] = 0$ for all $X \in \mathfrak{g}$.

(3) The *Jacobi identity*

$$[X, [Y, Z]] + [Y, [Z, X]] + [Z, [X, Y]] = 0$$

holds for all $X, Y, Z \in \mathfrak{g}$.

A subspace $\mathfrak{s}$ of $\mathfrak{g}$ is called a *subalgebra* if $[X, Y] \in \mathfrak{s}$ for all $X, Y \in \mathfrak{s}$; it is called an *ideal* if $[X, Y] \in \mathfrak{s}$ for all $X \in \mathfrak{g}$ and $Y \in \mathfrak{s}$.

An example of a Lie algebra is the *general linear algebra* $\mathfrak{gl}(V)$ consisting of all linear operators on a vector space V with the Lie bracket defined by

$$[X, Y] = XY - YX, \quad \forall X, Y \in \mathfrak{gl}(V).$$

The subspace $\mathfrak{sl}(V)$ consisting of all traceless linear operators is an ideal of $\mathfrak{gl}(V)$.

Let $\mathfrak{g}$ and $\mathfrak{h}$ be Lie algebras. A linear transformation $\varphi : \mathfrak{g} \to \mathfrak{h}$ is called a *homomorphism* if

$$\varphi([X, Y]) = [\varphi(X), \varphi(Y)], \quad \forall X, Y \in \mathfrak{g}.$$

It follows from the bilinearity and the Jacobi identity that the linear transformation
$$\mathrm{ad} : \mathfrak{g} \to \mathfrak{gl}(\mathfrak{g}),$$
defined by $\mathrm{ad}\, X(Y) = [X, Y]$ for all $X, Y \in \mathfrak{g}$, is a Lie algebra homomorphism, called the *adjoint representation* of $\mathfrak{g}$.

Structure Theory of Semisimple Lie Groups

A *Lie group* G is simultaneously a smooth manifold and a group such that the maps $m : G \times G \to G$ and $i : G \to G$ defined by group multiplication and group inversion are smooth.

The set of all nonsingular complex matrices forms a Lie group, called the *general linear group* and denoted by $\mathrm{GL}_n(\mathbb{C})$. Every closed subgroup of $\mathrm{GL}_n(\mathbb{C})$ is a Lie group, called a *closed linear group*.

Let G be a Lie group. For each $g \in G$, the left translation
$$L_g : G \to G,$$
defined by $L_g(h) = gh$ for all $h \in G$, is a diffeomorphism of G. A smooth vector field X on G is *left-invariant* if X is L_g-related to itself for every $g \in G$, i.e.,
$$X \circ L_g = dL_g \circ X, \qquad \forall\, g \in G.$$
If we regard X as a derivation on G, left-invariance is expressed by
$$(Xf) \circ L_g = X(f \circ L_g), \qquad \forall\, f \in C^\infty(G), \forall\, g \in G.$$
The space $\mathfrak{g}$ of left-invariant smooth vector fields on G is closed under the Lie bracket defined in Section 2.1, and is therefore a Lie algebra, called *the Lie algebra of G*.

There is another way to view the Lie algebra of a Lie group G. Let e denote the identity element of G. The map $X \mapsto X_e$ is a vector space isomorphism of the Lie algebra $\mathfrak{g}$ of G onto the tangent space $T_e(G)$ of G at e. If $X_e, Y_e \in T_e(G)$, let $[X_e, Y_e]$ denote the tangent vector $[X, Y]_e$. Then the vector space $T_e(G)$, with the composition rule $(X_e, Y_e) \mapsto [X_e, Y_e]$, forms a Lie algebra that is identified with $\mathfrak{g}$.

Let G and H be Lie groups. A smooth map $\varphi : G \to H$ is called a *smooth homomorphism* if it is also a group homomorphism. The differential
$$d\varphi : \mathfrak{g} \to \mathfrak{h}$$
between the corresponding Lie algebras $\mathfrak{g}$ and $\mathfrak{h}$ is a Lie algebra homomorphism, called the *derived homomorphism* of φ.

Let G be a Lie group with Lie algebra $\mathfrak{g}$. A *one-parameter subgroup* of G is a smooth homomorphism $\phi : \mathbb{R} \to G$. It is a consequence of the existence and uniqueness of solutions of linear ordinary differential equations that the map
$$\phi \mapsto d\phi(0)$$
is a bijection of the set of one-parameter subgroups of G onto $\mathfrak{g}$ [Hel78, p.103].

For each $X \in \mathfrak{g}$, let ϕ_X be the one-parameter subgroup corresponding to X. The *exponential map*
$$\exp : \mathfrak{g} \to G$$

is then defined by
$$\exp(X) = \phi_X(1), \quad \forall X \in \mathfrak{g}.$$
It follows that $\phi_X(t) = \exp(tX)$ for all $t \in \mathbb{R}$. Consequently, every one-parameter subgroup is of the form $t \mapsto \exp tX$ for some $X \in \mathfrak{g}$.

The exponential map for a closed linear group is given by the matrix exponential function [Kna02, p.76].

An important property of the exponential map is its naturality, i.e., if $\varphi : G \to H$ is a smooth homomorphism, then
$$\varphi \circ \exp_{\mathfrak{g}} = \exp_{\mathfrak{h}} \circ d\varphi. \tag{2.1}$$

A submanifold H of G is called a *Lie subgroup* if H is a Lie group with binary operation being the one induced by the binary operation on G. A Lie subgroup of G is called a *closed subgroup* if it is a closed subset of G. The following theorem establishes a one-to-one correspondence between connected Lie subgroups of a Lie group and subalgebras of its Lie algebra ([Hel78, p.112-113]).

Theorem 2.2. *Let G be a Lie group with Lie algebra $\mathfrak{g}$. If H is a Lie subgroup of G, then the Lie algebra $\mathfrak{h}$ of H is a subalgebra of $\mathfrak{g}$. Moreover,*
$$\mathfrak{h} = \{X \in \mathfrak{g} : \exp tX \in H \text{ for all } t \in \mathbb{R}\}.$$

Each subalgebra of $\mathfrak{g}$ is the Lie algebra of exactly one connected Lie subgroup of G.

For each $g \in G$, let I_g be the inner automorphism of G defined by
$$I_g(x) = gxg^{-1}, \quad \forall x \in G.$$

The derived homomorphism of I_g, denoted by $\mathrm{Ad}\, g$, is an automorphism of $\mathfrak{g}$. By the naturality of the exponential map (2.1), we thus have
$$\exp(Ad(g)X) = g(\exp X)g^{-1}, \quad \forall g \in G, \forall X \in \mathfrak{g}. \tag{2.2}$$

In the special case that G is a closed linear group with Lie algebra $\mathfrak{g}$, we have
$$\mathrm{Ad}\,(g)X = gXg^{-1}, \quad \forall g \in G, \forall X \in \mathfrak{g}.$$

Since the exponential map has a smooth inverse in a neighborhood of the identity $e \in G$, (2.2) implies that for each fixed small $X \in \mathfrak{g}$, the map $g \mapsto \mathrm{Ad}\,(g)X$ is smooth as a function from a neighborhood of e to $\mathfrak{g}$. In other words, $g \mapsto \mathrm{Ad}\, g$ is smooth from a neighborhood of e into $\mathrm{GL}(\mathfrak{g})$. Moreover,
$$\mathrm{Ad}\, g \circ \mathrm{Ad}\, h = \mathrm{Ad}\,(gh), \quad \forall g, h \in G,$$

since $I_g \circ I_h = I_{gh}$. Thus the smoothness is valid everywhere on G. Therefore

$$\mathrm{Ad} : G \to \mathrm{GL}(\mathfrak{g})$$

is a smooth homomorphism, called the *adjoint representation* of G. The derived homomorphism of Ad is the adjoint representation $\mathrm{ad} : \mathfrak{g} \to \mathfrak{gl}(\mathfrak{g})$ of $\mathfrak{g}$ [Kna02, p.80]. Consequently, by (2.1) we have

$$\mathrm{Ad}\,(\exp X) = \exp\,(\mathrm{ad}\, X), \qquad \forall X \in \mathfrak{g}.$$

The group $\mathrm{Aut}\,\mathfrak{g}$ of all automorphisms of $\mathfrak{g}$ is a closed subgroup of $\mathrm{GL}(\mathfrak{g})$, hence is a Lie subgroup of $\mathrm{GL}(\mathfrak{g})$. The Lie algebra of $\mathrm{Aut}\,\mathfrak{g}$, denoted by $\mathrm{Der}\,\mathfrak{g}$, consists of all derivations of $\mathfrak{g}$ [Hel78, p.127]. Since $\mathrm{ad}\,\mathfrak{g}$ is a subalgebra of $\mathrm{Der}\,\mathfrak{g}$, it corresponds to a connected subgroup $\mathrm{Int}\,\mathfrak{g}$ of $\mathrm{Aut}\,\mathfrak{g}$, which is generated by

$$\exp\,(\mathrm{ad}\,\mathfrak{g}) = \{\exp\,(\mathrm{ad}\,X) : X \in \mathfrak{g}\}$$

and called the *adjoint group* of $\mathfrak{g}$ [Hel78, p.127]. Since $\exp\,(\mathrm{ad}\,X) = \mathrm{Ad}\,(\exp X)$ for all $X \in \mathfrak{g}$, we have $\mathrm{Int}\,\mathfrak{g} = \mathrm{Ad}\,G$ if G is connected.

The Lie algebra $\mathfrak{g}$ is said to be *compact* if G is compact, or, equivalently, the adjoint group $\mathrm{Int}\,\mathfrak{g}$ is compact.

Let $\mathfrak{g}$ be a Lie algebra. The symmetric bilinear form B on $\mathfrak{g}$ defined by

$$B(X,Y) = \mathrm{tr}\,(\mathrm{ad}\,X\,\mathrm{ad}\,Y), \qquad \forall X, Y \in \mathfrak{g},$$

is called the *Killing form*, which is associative in the sense that

$$B([X,Y],Z) = B(X,[Y,Z]), \qquad \forall X, Y, Z \in \mathfrak{g}.$$

If σ is an automorphism of $\mathfrak{g}$, then

$$\mathrm{ad}\,(\sigma X) = \sigma \circ \mathrm{ad}\,X \circ \sigma^{-1},$$

and thus $B(\sigma X, \sigma Y) = B(X, Y)$. In particular, B is $\mathrm{Ad}\,G$-invariant.

A Lie algebra $\mathfrak{g}$ is *abelian* if $[\mathfrak{g},\mathfrak{g}] = 0$; it is *simple* if it is not abelian and has no nontrivial ideals; it is *solvable* if $D^k\mathfrak{g} = 0$ for some k, where $D^0\mathfrak{g} = \mathfrak{g}$ and $D^{k+1}\mathfrak{g} = [D^k\mathfrak{g}, D^k\mathfrak{g}]$; it is *nilpotent* if $C_k\mathfrak{g} = 0$ for some k, where $C_0\mathfrak{g} = \mathfrak{g}$ and $C_{k+1}\mathfrak{g} = [C_k\mathfrak{g},\mathfrak{g}]$; it is *semisimple* if its (unique) maximal solvable ideal, called the *radical* of $\mathfrak{g}$ and denoted by $\mathrm{Rad}\,\mathfrak{g}$, is trivial (or, equivalently, its Killing form is nondegenerate); it is *reductive* if its center $\mathfrak{z}(\mathfrak{g}) = \mathrm{Rad}\,\mathfrak{g}$ (or, equivalently, $[\mathfrak{g},\mathfrak{g}]$ is semisimple). A Lie algebra is semisimple if and only if it is isomorphic to a direct sum of simple algebras.

A Lie group is called *semisimple (simple, reductive, solvable, nilpotent, abelian)* if its Lie algebra is semisimple (simple, reductive, solvable, nilpotent, abelian).

2.3 Complex Semisimple Lie Algebras

Let $\mathfrak{g}$ be a complex semisimple Lie algebra. An element $X \in \mathfrak{g}$ is called *nilpotent* if $\operatorname{ad} X$ is a nilpotent endomorphism; it is called *semisimple* if $\operatorname{ad} X$ is diagonalizable. Since $\mathfrak{g}$ is semisimple, it possesses nonzero subalgebras consisting of semisimple elements, which are abelian and are called *toral subalgebras* of $\mathfrak{g}$ [Hum72, p.35].

The normalizer of a subalgebra $\mathfrak{a}$ of $\mathfrak{g}$ is

$$N_{\mathfrak{g}}(\mathfrak{a}) = \{X \in \mathfrak{g} : \operatorname{ad} X(\mathfrak{a}) \subset \mathfrak{a}\};$$

it is the largest subalgebra of $\mathfrak{g}$ which contains $\mathfrak{a}$ and in which $\mathfrak{a}$ is an ideal. A subalgebra $\mathfrak{h}$ of $\mathfrak{g}$ is called a *Cartan subalgebra* of $\mathfrak{g}$ if it is nilpotent and self-normalizing, i.e., $\mathfrak{h} = N_{\mathfrak{g}}(\mathfrak{h})$. The Cartan subalgebras of $\mathfrak{g}$ are exactly the maximal toral subalgebras of $\mathfrak{g}$ [Hum72, p.80]. All Cartan subalgebras of $\mathfrak{g}$ are conjugate under the adjoint group $\operatorname{Int} \mathfrak{g}$ of inner automorphisms [Hum72, p.82].

Let $\mathfrak{h}$ be a Cartan subalgebra of $\mathfrak{g}$. Since $\mathfrak{h}$ is abelian, $\operatorname{ad}_{\mathfrak{g}} \mathfrak{h}$ is a commuting family of semisimple endomorphisms of $\mathfrak{g}$, which are thus simultaneously diagonalizable. In other words, $\mathfrak{g}$ is the direct sum of the subspaces

$$\mathfrak{g}_{\alpha} = \{X \in \mathfrak{g} : [H, X] = \alpha(H)X \text{ for all } H \in \mathfrak{h}\},$$

where α ranges over the dual space $\mathfrak{h}^*$ of $\mathfrak{h}$. Note that $\mathfrak{g}_0 = \mathfrak{h}$ because $\mathfrak{h}$ is self-normalizing. A nonzero $\alpha \in \mathfrak{h}^*$ is called a *root* of $\mathfrak{g}$ relative to $\mathfrak{h}$ if $\mathfrak{g}_{\alpha} \neq 0$. The set of all roots, denoted by Δ, is called the *root system* of $\mathfrak{g}$ relative to $\mathfrak{h}$. Thus we have the *root space decomposition*

$$\mathfrak{g} = \mathfrak{h} \oplus \bigoplus_{\alpha \in \Delta} \mathfrak{g}_{\alpha}.$$

The importance of root space decomposition lies on the fact that Δ characterizes $\mathfrak{g}$ completely.

The restriction of the Killing form on $\mathfrak{h}$ is nondegenerate and is given by

$$B(H, H') = \sum_{\alpha \in \Delta} \alpha(H)\alpha(H'), \qquad \forall H, H' \in \mathfrak{h}.$$

Consequently, we can identify $\mathfrak{h}$ with $\mathfrak{h}^*$: each $\alpha \in \mathfrak{h}^*$ corresponds to a unique $H_{\alpha} \in \mathfrak{h}$ such that

$$\alpha(H) = B(H_{\alpha}, H), \qquad \forall H \in \mathfrak{h}.$$

And there is a nondegenerate bilinear form $\langle \cdot, \cdot \rangle$ defined on $\mathfrak{h}^*$ by

$$\langle \alpha, \beta \rangle = B(H_{\alpha}, H_{\beta}), \qquad \forall \alpha, \beta \in \mathfrak{h}^*.$$

The following is a collection of some properties of the root space decomposition [Hum72, p.36–40]:

(1) Δ is finite and spans $\mathfrak{h}^*$.

(2) If $\alpha, \beta \in \Delta \cup \{0\}$ and $\alpha + \beta \neq 0$, then $B(\mathfrak{g}_\alpha, \mathfrak{g}_\beta) = 0$.

(3) If $\alpha \in \Delta$, then $-\alpha \in \Delta$, but no other scalar multiple of α is a root.

(4) If $\alpha \in \Delta$, then $[\mathfrak{g}_\alpha, \mathfrak{g}_{-\alpha}]$ is one dimensional, with basis H_α.

(5) If $\alpha \in \Delta$, then $\dim \mathfrak{g}_\alpha = 1$.

(6) If $\alpha, \beta \in \Delta$, then $\dfrac{2\langle \beta, \alpha \rangle}{\langle \alpha, \alpha \rangle} \in \mathbb{Z}$ and $\beta - \dfrac{2\langle \beta, \alpha \rangle}{\langle \alpha, \alpha \rangle}\alpha \in \Delta$.

2.4 Real Forms

Let V be a vector space over $\mathbb{C}$. We can view V as a vector space $V_\mathbb{R}$ over $\mathbb{R}$, which is called the *realification* of V. To restore $V_\mathbb{R}$ to the complex vector space V, it suffices to know the $\mathbb{R}$-linear operator $J : V_\mathbb{R} \to V_\mathbb{R}$ defined by $J(X) = iX$ for all $X \in V_\mathbb{R}$. Note that $J^2 = -I$, where I is the identity operator. Then for any $X \in V_\mathbb{R}$ and for any complex number $a + bi$ with $a, b \in \mathbb{R}$, we have
$$(a + bi)X = av + bJ(X). \tag{2.3}$$

More generally, any linear operator J on a finite dimensional real vector space E satisfying $J^2 = -I$ is called a *complex structure* on E. Every finite dimensional vector space over $\mathbb{R}$ with a complex structure can be turned into a vector space over $\mathbb{C}$ by (2.3).

Let W be an arbitrary finite dimensional vector space over $\mathbb{R}$. The product $W \times W$ is again a vector space over $\mathbb{R}$. The linear operator J on $W \times W$ defined by $J : (X, Y) \mapsto (-Y, X)$ is a complex structure on $W \times W$, turning it into a complex vector space, which is denoted by $W_\mathbb{C}$ and called the *complexification* of W. Obviously, we have $\dim_\mathbb{C} W_\mathbb{C} = \dim_\mathbb{R} W$. Since for each $(X, Y) \in W_\mathbb{C}$, we have
$$(X, Y) = (X, 0) + J(Y, 0) = (X, 0) + i(Y, 0),$$
we write $X + iY$ instead of (X, Y).

On the other hand, every finite dimensional vector space V over $\mathbb{C}$ is isomorphic to $W_\mathbb{C}$ for some vector space W over $\mathbb{R}$, i.e., if $\{X_1, \ldots, X_n\}$ is any basis for V, then one may take $W = \left\{ \sum_{i=1}^n a_i X_i : a_i \in \mathbb{R} \right\}$.

Now let $\mathfrak{g}_0$ be a real Lie algebra. The complex vector space $\mathfrak{g} = (\mathfrak{g}_0)_\mathbb{C}$ then forms a Lie algebra over $\mathbb{C}$ with the Lie bracket defined by
$$[X + iY, A + iB] = ([X, A] - [Y, B]) + i([Y, A] + [X, B])$$

for all $X, Y, A, B \in \mathfrak{g}_0$. This complex Lie algebra $\mathfrak{g}$ is called the *complexification* of the real Lie algebra $\mathfrak{g}_0$.

For the remainder of this section, let $\mathfrak{g}$ be a complex Lie algebra.

The *realification* $\mathfrak{g}_\mathbb{R}$ of the complex vector space $\mathfrak{g}$ forms a real Lie algebra, with the Lie bracket inherited from $\mathfrak{g}$. A *real form* of $\mathfrak{g}$ is a subalgebra $\mathfrak{g}_0$ of $\mathfrak{g}_\mathbb{R}$ such that $\mathfrak{g}_\mathbb{R} = \mathfrak{g}_0 \oplus i\mathfrak{g}_0$. In this case, $\mathfrak{g}$ is isomorphic to the complexification of $\mathfrak{g}_0$.

Let $\mathfrak{g}_0$ be a real form of $\mathfrak{g}$. Each $Z \in \mathfrak{g}$ can be uniquely written as $Z = X + iY$ with $X, Y \in \mathfrak{g}_0$. A map $\sigma : \mathfrak{g} \to \mathfrak{g}$ given by $X + iY \mapsto X - iY$, where $X, Y \in \mathfrak{g}_0$, is called a *conjugation* of $\mathfrak{g}$ with respect to $\mathfrak{g}_0$. It is easy to see that

(1) $\sigma^2 = 1$,

(2) $\sigma(\alpha X) = \bar{\alpha}\sigma(X)$ for all $X \in \mathfrak{g}_0$ and $\alpha \in \mathbb{C}$,

(3) $\sigma(X + Y) = \sigma(X) + \sigma(Y)$ for all $X, Y \in \mathfrak{g}_0$, and

(4) $\sigma[X, Y] = [\sigma X, \sigma Y]$ for all $X, Y \in \mathfrak{g}_0$.

Thus σ is not an automorphism of $\mathfrak{g}$, but it is an automorphism of the real algebra $\mathfrak{g}_\mathbb{R}$.

On the other hand, if $\sigma : \mathfrak{g} \to \mathfrak{g}$ satisfies the above properties (1)–(4), the set $\mathfrak{g}_0$ of fixed points of σ is a real form of $\mathfrak{g}$ and σ is the conjugation of $\mathfrak{g}$ with respect to $\mathfrak{g}_0$.

Hence there is a one-to-one correspondence between real forms and conjugations of $\mathfrak{g}$.

A very important fact in the theory of semisimple Lie algebras is that every complex semisimple Lie algebra has a compact real form [Hel78, p.181]. The compact real forms of complex simple Lie algebra are listed in [Hel78, p.516].

We conclude this section with the following Killing form relations between a complex semisimple Lie algebra and its realification and real forms. Let B, $B_\mathbb{R}$, and B_0 denote the Killing forms of the Lie algebras $\mathfrak{g}$, $\mathfrak{g}_\mathbb{R}$, and $\mathfrak{g}_0$, respectively. Then

$$B_0(X, Y) = B(X, Y), \qquad \forall X, Y \in \mathfrak{g}_0$$
$$B_\mathbb{R}(X, Y) = 2\operatorname{Re} B(X, Y), \qquad \forall X, Y \in \mathfrak{g}_\mathbb{R}.$$

Consequently, $\mathfrak{g}$, $\mathfrak{g}_\mathbb{R}$, and $\mathfrak{g}_0$ are all semisimple if any of them is.

2.5 Cartan Decompositions

Let $\mathfrak{g}$ be a real semisimple Lie algebra, $\mathfrak{g}_\mathbb{C}$ its complexification, and σ the conjugation of $\mathfrak{g}_\mathbb{C}$ with respect to $\mathfrak{g}$. A direct sum decomposition

$$\mathfrak{g} = \mathfrak{k} \oplus \mathfrak{p}$$

of $\mathfrak{g}$ into a subalgebra $\mathfrak{k}$ and a vector subspace $\mathfrak{p}$ is called a *Cartan decomposition* if there exists a compact real form $\mathfrak{u}$ of $\mathfrak{g}_\mathbb{C}$ such that

$$\sigma(\mathfrak{u}) \subset \mathfrak{u}, \quad \mathfrak{k} = \mathfrak{g} \cap \mathfrak{u}, \quad \mathfrak{p} = \mathfrak{g} \cap i\mathfrak{u}.$$

If $\mathfrak{u}$ is any compact real form of $\mathfrak{g}_\mathbb{C}$ with a conjugation τ, then there exists an automorphism φ of $\mathfrak{g}_\mathbb{C}$ such that the compact real form $\varphi(\mathfrak{u})$ is invariant under σ, which guarantees the existence of a Cartan decomposition of $\mathfrak{g}$. In this case, the involutive automorphism $\theta = \sigma\tau$ is called a *Cartan involution* of $\mathfrak{g}$. This is equivalent to saying that the bilinear form B_θ of $\mathfrak{g}$ defined by

$$B_\theta(X, Y) = -B(X, \theta Y), \quad \forall X, Y \in \mathfrak{g},$$

is symmetric and strictly positive definite.

The following theorem establishes a one-to-one correspondence between Cartan decompositions of a real semisimple Lie algebra and its Cartan involutions (see [Hel78, p.184] and [OV94, p.144]).

Theorem 2.3. (Cartan Decomposition) *Let $\mathfrak{g}$ be a real semisimple Lie algebra with the direct sum of subspaces $\mathfrak{g} = \mathfrak{k} \oplus \mathfrak{p}$. Then the following statements are equivalent:*

(1) $\mathfrak{g} = \mathfrak{k} \oplus \mathfrak{p}$ is a Cartan decomposition.

(2) The map $\theta : X + Y \mapsto X - Y$, where $X \in \mathfrak{k}$ and $Y \in \mathfrak{p}$, is a Cartan involution of $\mathfrak{g}$.

(3) The Killing form B of $\mathfrak{g}$ is negative definite on $\mathfrak{k}$ and positive definite on $\mathfrak{p}$, and $[\mathfrak{k}, \mathfrak{k}] \subset \mathfrak{k}$, $[\mathfrak{p}, \mathfrak{p}] \subset \mathfrak{k}$, $[\mathfrak{k}, \mathfrak{p}] \subset \mathfrak{p}$.

Let $\mathfrak{g} = \mathfrak{k} \oplus \mathfrak{p}$ be a Cartan decomposition. Theorem 2.3 implies that $\mathfrak{k}$ and $\mathfrak{p}$ are the $+1$ and -1 eigenspaces of θ, respectively, and that $\mathfrak{k}$ is a maximal compactly embedded subalgebra of $\mathfrak{g}$. Moreover, $\mathfrak{k}$ and $\mathfrak{p}$ are orthogonal to each other with respect to both the Killing form B and the inner product B_θ.

For any $X \in \mathfrak{g}$, let $X_\mathfrak{k} \in \mathfrak{k}$ and $X_\mathfrak{p} \in \mathfrak{p}$ be the $\mathfrak{k}$-component and $\mathfrak{p}$-component of X, respectively, such that $X = X_\mathfrak{k} + X_\mathfrak{p}$.

Example 2.4. Consider the real simple Lie algebra $\mathfrak{g} = \mathfrak{sl}_n(\mathbb{R})$. Its complexification is $\mathfrak{g}_\mathbb{C} = \mathfrak{sl}_n(\mathbb{R}) \oplus i\mathfrak{sl}_n(\mathbb{R}) = \mathfrak{sl}_n(\mathbb{C})$. The conjugation σ of $\mathfrak{g}_\mathbb{C}$ corresponding to the real form $\mathfrak{g}$ is given by

$$\sigma(X + iY) = X - iY, \qquad \forall X, Y \in \mathfrak{g}.$$

Now $\mathfrak{g}_\mathbb{C}$ has a compact real form

$$\mathfrak{u} = \mathfrak{su}_n = \{A \in \mathfrak{sl}_n(\mathbb{C}) : A + A^* = 0\},$$

which is invariant under σ. The conjugation τ of $\mathfrak{g}_\mathbb{C}$ corresponding to the real form $\mathfrak{u}$ is given by

$$\tau(A) = -A^*, \qquad \forall A \in \mathfrak{g}_\mathbb{C}.$$

Thus

$$\mathfrak{k} = \mathfrak{g} \cap \mathfrak{u} = \mathfrak{sl}_n(\mathbb{R}) \cap \mathfrak{su}_n = \mathfrak{so}_n$$

and

$$\mathfrak{p} = \mathfrak{g} \cap i\mathfrak{u} = \mathfrak{sl}_n(\mathbb{R}) \cap i\mathfrak{su}_n = \{X : X \in \mathfrak{sl}_n(\mathbb{R}) \text{ is symmetric}\}.$$

The associated Cartan involution θ is given by

$$\theta(A) = \sigma(\tau(A)) = \sigma(-A^*) = \sigma(-A^\top) = -A^\top, \qquad \forall A \in \mathfrak{g}.$$

Moreover, up to a positive scalar, the bilinear form B_θ of $\mathfrak{g}$ is given by

$$B_\theta(X, Y) = -B(X, \theta Y) = B(X, Y^\top) = \operatorname{tr} XY^\top,$$

which is obviously an inner product on $\mathfrak{g} = \mathfrak{sl}_n(\mathbb{R})$.

In the special case of $\mathfrak{g}$ being a complex semisimple Lie algebra, if $\mathfrak{u}$ is a compact real form of $\mathfrak{g}$, then $\mathfrak{g}_\mathbb{R} = \mathfrak{u} \oplus i\mathfrak{u}$ is a Cartan decomposition [Hel78, p.185].

Example 2.5. Consider the complex simple Lie algebra $\mathfrak{g} = \mathfrak{sl}_n(\mathbb{C})$, viewed as a real Lie algebra $\mathfrak{g}_\mathbb{R}$. Now that $\mathfrak{su}_n$ is a compact real form of $\mathfrak{g}$, we see that

$$\mathfrak{sl}_n(\mathbb{C}) = \mathfrak{su}_n \oplus i\mathfrak{su}_n$$

is a Cartan decomposition of $\mathfrak{g}_\mathbb{R}$, with $\mathfrak{k} = \mathfrak{su}_n$ consisting of skew-Hermitian matrices in $\mathfrak{g}$ and $\mathfrak{p} = i\mathfrak{su}_n$ consisting of Hermitian matrices in $\mathfrak{g}$. The corresponding Cartan involution θ is given by

$$\theta(A) = -A^*, \qquad \forall A \in \mathfrak{sl}_n(\mathbb{C}),$$

and the symmetric positive definite bilinear form B_θ is given by (up to a positive scalar)

$$B_\theta(X, Y) = \operatorname{tr} XY^*, \qquad \forall X, Y \in \mathfrak{sl}_n(\mathbb{C}).$$

Structure Theory of Semisimple Lie Groups 53

The group level *Cartan decomposition* is summarized below (see [Hel78, p.252] and [Kna02, p.362]).

Theorem 2.6. (Cartan Decomposition) *Let G be a real noncompact semisimple Lie group with Lie algebra $\mathfrak{g}$. Let $\mathfrak{g} = \mathfrak{k} \oplus \mathfrak{p}$ be the Cartan decomposition corresponding to a Cartan involution θ of $\mathfrak{g}$. Let K be the analytic subgroup of G with Lie algebra $\mathfrak{k}$. Then*

(1) *K is connected, closed, and contains the center Z of G. Moreover, K is compact if and only if Z is finite; in this case, K is a maximal compact subgroup of G.*

(2) *There exists an involutive, analytic automorphism Θ of G whose fixed point set is K and whose differential at the identity of G is θ.*

(3) *The maps $K \times \mathfrak{p} \to G$ given by $(k, X) \mapsto k(\exp X)$ and $\mathfrak{p} \times K \to G$ given by $(X, k) \mapsto (\exp X)k$, respectively, are diffeomorphisms onto G.*

The automorphism $\Theta : G \to G$ is called a *Cartan involution* of G.

Let $P = \exp \mathfrak{p} = \{\exp X : X \in \mathfrak{p}\}$. By Theorem 2.6, the exponential map $\exp_\mathfrak{g} : \mathfrak{p} \to P$ is a diffeomorphism onto P. So for each $p \in P$, there exists a unique $X \in \mathfrak{p}$ such that $p = \exp X$. By the naturality of the exponential map, we have

$$\Theta(p) = p^{-1}, \qquad \forall p \in P.$$

For any $k \in K$, $\operatorname{Ad} k$ leaves B invariant because $\operatorname{Ad} k \in \operatorname{Aut} \mathfrak{g}$; $\operatorname{Ad} k$ also leaves $\mathfrak{k}$ invariant because $\mathfrak{k}$ is the Lie algebra of K, and hence $\operatorname{Ad} k$ leaves invariant the subspace of $\mathfrak{g}$ orthogonal to $\mathfrak{k}$, which is exactly $\mathfrak{p}$. If $X \in \mathfrak{g}$, write $X = X_\mathfrak{k} + X_\mathfrak{p}$ with $X_\mathfrak{k} \in \mathfrak{k}$ and $X_\mathfrak{p} \in \mathfrak{p}$ and we see that

$$\begin{aligned}\operatorname{Ad} k(\theta(X)) &= \operatorname{Ad}(k)X_\mathfrak{k} - \operatorname{Ad}(k)X_\mathfrak{p} \\ &= \theta(\operatorname{Ad}(k)X_\mathfrak{k}) + \theta(\operatorname{Ad}(k)X_\mathfrak{p}) \\ &= \theta(\operatorname{Ad}(k)X),\end{aligned}$$

i.e., $\operatorname{Ad} k$ commutes with θ. Hence $\operatorname{Ad} k$ leaves B_θ invariant as well.

Example 2.7. Consider the real simple Lie group $G = \operatorname{SL}_n(\mathbb{C})$, whose Lie algebra is $\mathfrak{g} = \mathfrak{sl}_n(\mathbb{C})$. The center of G is

$$Z = \begin{cases} \{I\} & \text{if } n \text{ is odd;} \\ \{I, -I\} & \text{if } n \text{ is even.} \end{cases}$$

As in Example 2.5, $\mathfrak{g} = \mathfrak{su}_n \oplus i\mathfrak{su}_n$ is a Cartan decomposition with the Cartan involution $\theta : A \mapsto -A^*$. Then $K = \operatorname{SU}_n$, which is connected and compact and contains the finite center Z. The involutive automorphism Θ of G is given by

$$\Theta(g) = (g^{-1})^*, \qquad \forall g \in G. \tag{2.4}$$

Obviously, $d\Theta = \theta$ and the fixed point set of Θ is K. Finally, if we denote

$$P = \exp\mathfrak{p} = \exp(i\mathfrak{su}_n),$$

the group level Cartan decompositions are $G = PK$ and $G = KP$, which correspond to the usual left and right polar decompositions for matrices on the element level, respectively.

Inspired by (2.4) in Example 2.7, we define a diffeomorphism $* : G \to G$ by

$$*(g) = \Theta(g^{-1}), \qquad \forall\, g \in G. \tag{2.5}$$

We also write $g^* = *(g)$ for convenience. Note that $*$ is not an automorphism, because

$$(fg)^* = g^* f^*, \qquad \forall\, f, g \in G.$$

It is easy to see that

$$k^* = k^{-1}, \qquad \forall\, k \in K,$$

and that

$$p^* = p, \qquad \forall\, p \in P.$$

Moreover, if $g = kp$ is the Cartan decomposition of g (with $k \in K$ and $p \in P$), then

$$g^* = pk^{-1} \quad\text{and}\quad g^*g = p^2 \in P. \tag{2.6}$$

Now we define normal elements in G. An element $g \in G$ is said to be *normal* if

$$gg^* = g^*g. \tag{2.7}$$

As in [Hel78, p.183], Cartan decomposition is unique up to conjugation, so normality is independent of the choice of K and P (and hence Θ). Obviously, elements in P are normal. If $g = kp$ is the Cartan decomposition of $g \in G$, then g is normal if and only if $kp = pk$. By Example 2.7, we see that normality in G is reduced to the usual normality of matrices for $G = \mathrm{SL}_n(\mathbb{C})$.

Since θ is the differential of Θ at the identity, by the naturality of the exponential map, we have

$$(e^X)^* = \Theta(e^{-X}) = e^{-\theta X}, \qquad \forall\, X \in \mathfrak{g}.$$

Thus the differential of $*$, also denoted by $*$, is just $-\theta$. Similar to the group case, we denote $*(X) = X^*$ for all $X \in \mathfrak{g}$. Thus

$$(X^*)^* = X \quad\text{and}\quad X^* + X \in \mathfrak{p}, \qquad \forall\, X \in \mathfrak{g}.$$

Moreover, $\mathfrak{p}$ is the eigenspace of $* : \mathfrak{g} \to \mathfrak{g}$ associated with the eigenvalue 1 and $\mathfrak{k}$ is the eigenspace of $*$ associated with the eigenvalue -1. Consequently, P is the fixed point set of $* : G \to G$.

An element $X \in \mathfrak{g}$ is said to be *normal* if $[X^*, X] = 0$. Obviously, if $X \in \mathfrak{g}$ is normal, then e^X is normal in G.

2.6 Root Space Decomposition

Let $\mathfrak{g}$ be a real semisimple Lie algebra, and let $\mathfrak{g} = \mathfrak{k} \oplus \mathfrak{p}$ be a Cartan decomposition with θ the corresponding Cartan involution. The bilinear form B_θ endows $\mathfrak{g}$ with the structure of a finite-dimensional inner product space. For any $X \in \mathfrak{g}$, with respect to B_θ, the adjoint of $\operatorname{ad} X$ is $-\operatorname{ad}\theta(X)$. This is because for all $Y, Z \in \mathfrak{g}$, we have

$$\begin{aligned}
B_\theta((\operatorname{ad}\theta X)Y, Z) &= -B([\theta X, Y], \theta Z) \\
&= B(Y, [\theta X, \theta Z]) \\
&= B(Y, \theta[X, Z]) \\
&= -B_\theta(Y, (\operatorname{ad} X)Z) \\
&= -B_\theta((\operatorname{ad} X)^* Y, Z),
\end{aligned}$$

where $(\operatorname{ad} X)^*$ denotes the adjoint of $\operatorname{ad} X$. If $X \in \mathfrak{p}$, then $\theta(X) = -X$ and hence $\operatorname{ad} X$ is self-adjoint, which means $\operatorname{ad} X$ can be represented by a symmetric matrix with respect to an orthonormal basis of $\mathfrak{g}$. Thus the elements of $\operatorname{ad} \mathfrak{p}$ are semisimple with real eigenvalues.

Let $\mathfrak{a}$ be a maximal abelian subspace of $\mathfrak{p}$. The commutative family $\operatorname{ad}\mathfrak{a}$ is simultaneously diagonalizable. For each real linear functional α on $\mathfrak{a}$, let

$$\mathfrak{g}_\alpha = \{ X \in \mathfrak{g} : [H, X] = \alpha(H)X \text{ for all } H \in \mathfrak{a} \}.$$

It is easy to see that $\theta(\mathfrak{g}_\alpha) = \mathfrak{g}_{-\alpha}$ and $[\mathfrak{g}_\alpha, \mathfrak{g}_\beta] \subset \mathfrak{g}_{\alpha+\beta}$. Moreover, if $\mathfrak{m}$ is the centralizer of $\mathfrak{a}$ in $\mathfrak{k}$, i.e., $\mathfrak{m} = \{X \in \mathfrak{k} : \operatorname{ad}(X)H = 0 \text{ for all } H \in \mathfrak{a}\}$, then by the maximality of $\mathfrak{a}$ as an abelian subspace of $\mathfrak{p}$,

$$\mathfrak{g}_0 = (\mathfrak{g}_0 \cap \mathfrak{k}) \oplus (\mathfrak{g}_0 \cap \mathfrak{p}) = \mathfrak{m} \oplus \mathfrak{a}$$

is an orthogonal decomposition. If $\alpha \neq 0$ and $\mathfrak{g}_\alpha \neq \{0\}$, then α is called a *root* of $\mathfrak{g}$ with respect to $\mathfrak{a}$.

Let Σ denote the set of all roots, which is obviously finite. The simultaneous diagonalization of $\operatorname{ad}\mathfrak{a}$ is then expressed by the orthogonal direct sum

$$\mathfrak{g} = \mathfrak{g}_0 \oplus \bigoplus_{\alpha \in \Sigma} \mathfrak{g}_\alpha, \tag{2.8}$$

which is called the *restricted root space decomposition* of $\mathfrak{g}$ with respect to $\mathfrak{a}$.

For each $\alpha \in \Sigma$, the set

$$P_\alpha = \{X \in \mathfrak{a} : \alpha(X) = 0\}$$

is a subspace of $\mathfrak{a}$ of codimension 1. The subspaces P_α for all $\alpha \in \Sigma$ divide

$\mathfrak{a}$ into several open convex cones, called *Weyl chambers*. Fix a Weyl chamber $\mathfrak{a}_+$ and refer to it as the *fundamental Weyl chamber*. A root α is *positive* if it is positive on $\mathfrak{a}_+$. Let Σ^+ denote the set of all positive roots.

Suppose $\alpha \in \Sigma$ and $X \in \mathfrak{g}_\alpha$ and $H \in \mathfrak{a}$. Write $X = X_\mathfrak{k} + X_\mathfrak{p}$ with $X_\mathfrak{k} \in \mathfrak{k}$ and $X_\mathfrak{p} \in \mathfrak{p}$. Since $[\mathfrak{k}, \mathfrak{p}] \subset \mathfrak{p}$ and $[\mathfrak{p}, \mathfrak{p}] \subset \mathfrak{k}$, we have $[H, X_\mathfrak{k}] \in \mathfrak{p}$ and $[H, X_\mathfrak{p}] \in \mathfrak{k}$. Note that

$$\alpha(H)X_\mathfrak{k} + \alpha(H)X_\mathfrak{p} = \alpha(H)X = [H, X] = [H, X_\mathfrak{p}] + [H, X_\mathfrak{k}].$$

Thus we have $(\operatorname{ad} H)X_\mathfrak{k} = \alpha(H)X_\mathfrak{p}$ and $(\operatorname{ad} H)X_\mathfrak{p} = \alpha(H)X_\mathfrak{k}$, which imply that

$$(\operatorname{ad} H)^2 X_\mathfrak{k} = \alpha(H)^2 X_\mathfrak{k} \quad \text{and} \quad (\operatorname{ad} H)^2 X_\mathfrak{p} = \alpha(H)^2 X_\mathfrak{p}.$$

Now for each $\alpha \in \Sigma$, the following are well-defined:

$$\mathfrak{k}_\alpha = \{X \in \mathfrak{k} : (\operatorname{ad} H)^2 X = \alpha(H)^2 X \text{ for all } H \in \mathfrak{a}\},$$
$$\mathfrak{p}_\alpha = \{X \in \mathfrak{p} : (\operatorname{ad} H)^2 X = \alpha(H)^2 X \text{ for all } H \in \mathfrak{a}\}.$$

So $\mathfrak{k}_\alpha = \mathfrak{k}_{-\alpha}$ and $\mathfrak{p}_\alpha = \mathfrak{p}_{-\alpha}$ for all $\alpha \in \Sigma$.

Then we have the following result (see [Lia04, p.107]).

Theorem 2.8. *The following statements are true.*

(1) $\mathfrak{k} = \mathfrak{m} \oplus \bigoplus_{\alpha \in \Sigma^+} \mathfrak{k}_\alpha$ *and* $\mathfrak{p} = \mathfrak{a} \oplus \bigoplus_{\alpha \in \Sigma^+} \mathfrak{p}_\alpha$ *are direct sums whose components are mutually orthogonal under* B_θ.

(2) $\mathfrak{g}_\alpha \oplus \mathfrak{g}_{-\alpha} = \mathfrak{k}_\alpha \oplus \mathfrak{p}_\alpha$ *for all* $\alpha \in \Sigma^+$.

(3) $\dim \mathfrak{g}_\alpha = \dim \mathfrak{k}_\alpha = \dim \mathfrak{p}_\alpha$ *for all* $\alpha \in \Sigma^+$.

Example 2.9. Consider a root space decomposition for the real simple Lie algebra $\mathfrak{g} = \mathfrak{sl}_n(\mathbb{C})$. As in Example 2.5, let

$$\mathfrak{sl}_n(\mathbb{C}) = \mathfrak{su}_n \oplus i\mathfrak{su}_n$$

be a Cartan decomposition with corresponding Cartan involution given by

$$\theta(X) = -X^*, \quad \forall X \in \mathfrak{g},$$

and let the inner product B_θ on $\mathfrak{g}$ be given by

$$B_\theta(X, Y) = \operatorname{tr} XY^*, \quad \forall X, Y \in \mathfrak{g}.$$

Let $\mathfrak{a} \subset \mathfrak{p}$ be the set of all real diagonal matrices with trace zero, which is a maximal abelian subspace of $\mathfrak{p}$. For each $1 \leqslant i \leqslant n$, let $f_i : \mathfrak{a} \to \mathbb{R}$ be evaluation of the ith diagonal entry. Then the set of restricted roots of $\mathfrak{g}$ with respect to $\mathfrak{a}$ is

$$\Sigma = \{f_i - f_j : 1 \leqslant i \neq j \leqslant n\}$$

and
$$\mathfrak{g}_{f_i-f_j} = \mathbb{C} E_{ij},$$
where $\{E_{ij} : 1 \leqslant i \leqslant n, 1 \leqslant j \leqslant n\}$ is the standard basis for $\mathbb{C}_{n \times n}$. Moreover,
$$\mathfrak{g}_0 = \mathfrak{m} + \mathfrak{a},$$
where $\mathfrak{m} = i\mathfrak{a}$ consists of all purely imaginary matrices with trace zero. Choose the open Weyl chamber
$$\mathfrak{a}_+ = \{H = \mathrm{diag}\,(h_1, \ldots, h_n) \in \mathfrak{a} : h_1 > \cdots > h_n\}$$
as the fundamental Weyl chamber. Then the set of positive roots is
$$\Sigma_+ = \{f_i - f_j : 1 \leqslant i < j \leqslant n\}.$$
For each $\alpha = f_i - f_j \in \Sigma_+$,
$$\mathfrak{k}_\alpha = \mathbb{R}(E_{ij} - E_{ji}) + \sqrt{-1}\mathbb{R}(E_{ij} + E_{ji}),$$
$$\mathfrak{p}_\alpha = \mathbb{R}(E_{ij} + E_{ji}) + \sqrt{-1}\mathbb{R}(E_{ij} - E_{ji}).$$

2.7 Iwasawa Decompositions

Now we turn to another important decomposition of $\mathfrak{g}$. Let the notations be as in Section 2.6. The space
$$\mathfrak{n} = \bigoplus_{\alpha \in \Sigma^+} \mathfrak{g}_\alpha$$
is a subalgebra of $\mathfrak{g}$. Since $\theta \mathfrak{g}_\alpha = \mathfrak{g}_{-\alpha}$ for all $\alpha \in \Sigma$, we have
$$\mathfrak{g} = \mathfrak{g}_0 \oplus \mathfrak{n} \oplus \theta \mathfrak{n},$$
according to the root space decomposition (2.8). If $X \in \bigoplus_{\alpha \in \Sigma^+} \mathfrak{g}_{-\alpha}$, then
$$X = (X + \theta(X)) - \theta(X) \in \mathfrak{k} + \mathfrak{n}.$$
Because $\mathfrak{g}_0 = \mathfrak{m} \oplus \mathfrak{a}$, the root space decomposition (2.8), together with the above observations, yields that
$$\mathfrak{g} = \mathfrak{k} + \mathfrak{a} + \mathfrak{n}.$$
Obviously, $\mathfrak{a} \cap \mathfrak{n} = \{0\}$. Now suppose $X \in \mathfrak{k} \cap (\mathfrak{a} \oplus \mathfrak{n})$. Then
$$\mathfrak{a} \oplus \mathfrak{n} \ni X = \theta X \in \mathfrak{a} \oplus \theta \mathfrak{n},$$

which then implies that $X \in \mathfrak{a}$. But $\mathfrak{k} \cap \mathfrak{a} = \{0\}$, we then conclude that $X = 0$. This shows that $\mathfrak{g} = \mathfrak{k} + \mathfrak{a} + \mathfrak{n}$ is a direct sum. In other words, we have

$$\mathfrak{g} = \mathfrak{k} \oplus \mathfrak{a} \oplus \mathfrak{n}, \tag{2.9}$$

which is called an *Iwasawa decomposition* of $\mathfrak{g}$ (see [Hel78, p.263] and [Kna02, p.373]).

Example 2.10. Consider an Iwasawa decomposition for the real simple Lie algebra $\mathfrak{g} = \mathfrak{sl}_n(\mathbb{C})$. Let the Cartan decomposition and root space decomposition of $\mathfrak{g}$ be as in Example 2.9. So $\mathfrak{k} = \mathfrak{su}_n$ consists of skew-Hermitian matrices in $\mathfrak{sl}_n(\mathbb{C})$, $\mathfrak{a}$ consists of real diagonal matrices, and $\mathfrak{n}$ consists of strictly upper triangular matrices. The Iwasawa decomposition $\mathfrak{g} = \mathfrak{k} \oplus \mathfrak{a} \oplus \mathfrak{n}$ is then the one as in (1.7).

The following theorem summarizes the Iwasawa decomposition on the group level (see [Kna02, p.374]).

Theorem 2.11. (Iwasawa Decomposition) *Let G be a real noncompact semisimple Lie group with Lie algebra $\mathfrak{g}$. Let $\mathfrak{g} = \mathfrak{k} \oplus \mathfrak{a} \oplus \mathfrak{n}$ be an Iwasawa decomposition. Let K, A, and N be the analytic subgroups of G with Lie algebras $\mathfrak{k}$, $\mathfrak{a}$, and $\mathfrak{n}$, respectively. Then*

$$G = KAN$$

and the map $(k, a, n) \mapsto kan$ is a diffeomorphism of $K \times A \times N$ onto G.

Example 2.12. Consider an Iwasawa decomposition for the real simple Lie group $G = \mathrm{SL}_n(\mathbb{C})$ whose Lie algebra is $\mathfrak{g} = \mathfrak{sl}_n(\mathbb{C})$. Let $\mathfrak{g} = \mathfrak{k} \oplus \mathfrak{a} \oplus \mathfrak{n}$ be the Iwasawa decomposition as in Example 2.10. Then $K = \exp \mathfrak{su}_n = \mathrm{SU}_n$, $A = \exp \mathfrak{a}$ is the subgroup of G of diagonal matrices with positive diagonal entries, and $N = \exp \mathfrak{n}$ is the subgroup of G of upper triangular matrices whose diagonal entries are all 1. The decomposition $G = KAN$ amounts to the Gram-Schmidt orthogonalization process. To see this, let $g \in G$ be given and let $\{e_1, \ldots, e_n\}$ be the standard basis for $\mathbb{C}^n$. Then $\{ge_1, \ldots, ge_n\}$ forms another basis of $\mathbb{C}^n$. The Gram-Schmidt orthogonalization process yields an orthonormal basis $\{u_1, \ldots, u_n\}$ such that for all $1 \leqslant i \leqslant n$,

$$\mathrm{span}\{ge_1, \ldots, ge_j\} = \mathrm{span}\{u_1, \ldots, u_i\}$$

and

$$u_i \in \mathbb{R}^+(ge_i) + \mathrm{span}\{u_1, \ldots, u_{i-1}\}.$$

Let $k \in \mathrm{U}_n$ be such that $ke_i = u_i$ for all $1 \leqslant i \leqslant n$. Then the above Gram-Schmidt process implies that $k^{-1}g$ is upper triangular with positive diagonal entries. In other words, $k^{-1}g \in AN$. Since $k \in \mathrm{U}_n$ and

$$\det k^{-1} = \det k^{-1} \det g = \det(k^{-1}g) > 0,$$

we see that $k \in \mathrm{SU}_n = K$. Thus

$$g = k(k^{-1}g) \in KAN$$

is the QR decomposition as in (1.4).

2.8 Weyl Groups

Let the notations be as in Section 2.6 and Section 2.7. Let $\mathfrak{m}$ and M be the centralizers of $\mathfrak{a}$ in $\mathfrak{k}$ and in K, respectively, and M' the normalizer of $\mathfrak{a}$ in K, i.e.,

$$\mathfrak{m} = \{X \in \mathfrak{k} : \mathrm{ad}\,(X)H = 0 \text{ for all } H \in \mathfrak{a}\},$$
$$M = \{k \in K : \mathrm{Ad}\,(k)H = H \text{ for all } H \in \mathfrak{a}\},$$
$$M' = \{k \in K : \mathrm{Ad}\,(k)\mathfrak{a} \subset \mathfrak{a}\}.$$

Let $A = \exp \mathfrak{a}$ be the analytical subgroup of G with Lie algebra $\mathfrak{a}$. Note that M and M' are also the centralizer and normalizer of A in K, respectively, and that they are closed Lie subgroups of K. More importantly, M is a normal subgroup of M', and the quotient group M'/M is finite, because M and M' have the same Lie algebra $\mathfrak{m}$ (see [Hel78, p.284]).

The finite group $W(G, A) = M'/M$ is called the (analytically defined) *Weyl group* of G relative to A. For $w = m_w M \in W(G, A)$, the linear map $\mathrm{Ad}\,(m_w) : \mathfrak{a} \to \mathfrak{a}$ does not depend on the choice of $m_w \in M'$ representing w. Therefore, $w \mapsto \mathrm{Ad}\,(m_w)$ is a faithful representation of $W(G, A)$ on $\mathfrak{a}$. Thus we may regard $w \in W(G, A)$ as the linear map $\mathrm{Ad}\,(m_w) : \mathfrak{a} \to \mathfrak{a}$ and $W(G, A)$ as a group of linear operators on $\mathfrak{a}$.

The Killing form B is nondegenerate on $\mathfrak{a}$, and thus it induces an isomorphism of $\mathfrak{a}^*$ and $\mathfrak{a}$ by $\lambda \mapsto H_\lambda$ such that

$$\lambda(H) = B(H_\lambda, H), \qquad \forall \lambda \in \mathfrak{a}^*, \forall H \in \mathfrak{a}.$$

This isomorphism induces an action of $W(G, A)$ on $\mathfrak{a}^*$ as follows. If we denote $H_{w \cdot \lambda} = w \cdot H_\lambda$ for all $\lambda \in \mathfrak{a}^*$, then for all $H \in \mathfrak{a}$,

$$\begin{aligned}(w \cdot \lambda)(H) &= B(H_{w \cdot \lambda}, H) \\ &= B(w \cdot H_\lambda, H) \\ &= B(\mathrm{Ad}\,(k_w)H_\lambda, H) \\ &= B(H_\lambda, \mathrm{Ad}\,(k_w)^{-1}H) \\ &= \lambda(\mathrm{Ad}\,(k_w)^{-1}H) \\ &= (\lambda \circ \mathrm{Ad}\,(k_w)^{-1})(H).\end{aligned}$$

So the Weyl group $W(G, A)$ acts on $\mathfrak{a}^*$ by

$$w \cdot \lambda = \lambda \circ \mathrm{Ad}\,(k_w)^{-1} := \lambda \circ w^{-1}, \qquad \forall \lambda \in \mathfrak{a}^*.$$

The Weyl group can also be defined in an algebraic approach. For each root $\alpha \in \Sigma$, the reflection s_α about the hyperplane $P_\alpha = \{X \in \mathfrak{a} : \alpha(X) = 0\}$, with respect to the Killing form B, is a linear map on $\mathfrak{a}$ given by

$$s_\alpha(H) = H - \frac{2\alpha(H)}{\alpha(H_\alpha)} H_\alpha, \quad \forall H \in \mathfrak{a},$$

where H_α is the element of $\mathfrak{a}$ representing α, i.e., $\alpha(H) = B(H, H_\alpha)$ for all $H \in \mathfrak{a}$. The group $W(\mathfrak{g}, \mathfrak{a})$ generated by $\{s_\alpha : \alpha \in \Sigma\}$ is called the (algebraically defined) *Weyl group* of $\mathfrak{g}$ relative to $\mathfrak{a}$. When viewed as groups of linear operators on $\mathfrak{a}$, the two Weyl groups $W(G, A)$ and $W(\mathfrak{g}, \mathfrak{a})$ coincide (see [Kna02, p.383]).

Example 2.13. Let $G = \mathrm{SL}_n(\mathbb{C})$ and $\mathfrak{g} = \mathfrak{sl}_n(\mathbb{C})$. Let the Cartan decomposition and root space decomposition of $\mathfrak{g}$ be as in Examples 2.5 and 2.9. Let $\mathfrak{a}$ be the maximal subspace of $\mathfrak{p}$ consisting of real diagonal matrices $\mathfrak{g}$. Since $\mathfrak{k} = \mathfrak{su}_n$ and $K = \mathrm{SU}_n$, we have

$$\mathfrak{m} = i\mathfrak{a},$$
$$M = \{k \in K : k \text{ is diagonal}\},$$
$$M' = \{k \in K : k \text{ is a generalized permutation matrix}\}$$
$$= M \rtimes \mathbb{S}_n,$$

where $M \rtimes \mathbb{S}_n$ denotes the semidirect product M and $\mathbb{S}_n$, with M the normal subgroup. Thus the Weyl group $W(G, A) = M'/M \cong \mathbb{S}_n$.

2.9 KA_+K Decomposition

In this section, let the notations be as in Section 2.5 and Section 2.7. More precisely, let G be a real noncompact semisimple Lie group with Lie algebra $\mathfrak{g}$, let $\mathfrak{g} = \mathfrak{k} \oplus \mathfrak{p}$ be a Cartan decomposition corresponding to a Cartan involution θ, let K be the analytic subgroup of G with Lie algebra $\mathfrak{k}$, and let $P = \exp \mathfrak{p}$ so that the exponential map $\exp_\mathfrak{g} : \mathfrak{p} \to P$ is a diffeomorphism onto P. Let $\mathfrak{a}$ be a maximal abelian subspace of $\mathfrak{p}$. Fix a closed Weyl chamber $\mathfrak{a}_+$ of $\mathfrak{a}$. Let $A_+ = \exp \mathfrak{a}_+$. Every element in $\mathfrak{p}$ is K-conjugate to a unique element in $\mathfrak{a}_+$ (see [Kna02, p.378]). In other words, if $X \in \mathfrak{p}$, there exist a unique $Z \in \mathfrak{a}_+$ and some $k \in K$ such that
$$X = \mathrm{Ad}\, k(Z);$$
it follows that
$$\exp X = \exp(\mathrm{Ad}\, k(Z)) = k \exp(Z) k^{-1} \in KA_+K. \qquad (2.10)$$

Applying (2.10) to the P-component of any $g \in G$ with Cartan decompositions $g = kp$ or $g = p'k'$, we have the following Lie group decomposition as an extension of singular value decomposition (see [Hel78, p.402]).

Theorem 2.14. (KA_+K Decomposition) *Let G be any noncompact connected semisimple Lie group. Then*

$$G = KA_+K. \tag{2.11}$$

In other words, each $g \in G$ can be written as

$$g = uav, \tag{2.12}$$

where $u, v \in K$ and $a \in A_+$ is uniquely determined.

Example 2.15. Consider the real simple Lie group $G = \mathrm{SL}_n(\mathbb{C})$. Let the Cartan decompositions $G = KP$ and $G = PK$ be as in Example 2.7. Let $\mathfrak{a}_+$ be the closed fundamental Weyl chamber of $\mathfrak{a}$ consisting of all real diagonal matrices whose diagonal entries are in decreasing order. Then A_+ is the set of all real diagonal matrices whose diagonal entries are positive and in decreasing order. The group decomposition $G = KA_+K$ is then the usual singular value decomposition for $\mathrm{SL}_n(\mathbb{C})$.

2.10 Complete Multiplicative Jordan Decomposition

In this section, let the notations be as in Section 2.5 and Section 2.7 and Section 2.8. Let G be a real noncompact connected semisimple Lie group with Lie algebra $\mathfrak{g}$. Let $\mathfrak{g} = \mathfrak{k} \oplus \mathfrak{p}$ be a fixed Cartan decomposition of $\mathfrak{g}$, with θ the corresponding Cartan involution. Let Θ be the derived Cartan involution of G. Let

$$G = PK \quad \text{and} \quad G = KP$$

be the left and right Cartan decompositions, respectively. Note that if $p = \exp X$ with $X \in \mathfrak{p}$, then

$$p^{-1} = \exp(-X) \in P, \quad p^2 = \exp(2X) \in P, \quad p^{1/2} = \exp(X/2) \in P.$$

Pick a maximal abelian subspace $\mathfrak{a}$ of $\mathfrak{p}$. Let $A = \exp \mathfrak{a}$ be the analytic subgroup of G generated by $\mathfrak{a}$. It is known [Kna02, p.378] that

$$\mathfrak{p} = \mathrm{Ad}\, K(\mathfrak{a}).$$

The Weyl group W of $(\mathfrak{g}, \mathfrak{a})$ acts simply transitively on $\mathfrak{a}$, and also on A through the exponential map $\exp : \mathfrak{a} \to A$. Let $\mathfrak{g} = \mathfrak{k} \oplus \mathfrak{a} \oplus \mathfrak{n}$ and $G = KAN$ be the corresponding Iwasawa decompositions.

An element $X \in \mathfrak{g}$ is called *real semisimple* (resp., *nilpotent*) if $\mathrm{ad}\, X$ is diagonalizable over $\mathbb{R}$ (resp., $\mathrm{ad}\, X$ is nilpotent).

An element $g \in G$ is called *hyperbolic* (resp., *unipotent*) if $g = \exp X$ for some real semisimple (resp., nilpotent) $X \in \mathfrak{g}$; in either case, X is unique and we write $X = \log g$.

An element $g \in G$ is called *elliptic* if $\operatorname{Ad} g$ is diagonalizable over $\mathbb{C}$ with eigenvalues of modulus 1.

The following result characterizes elliptic, hyperbolic, and unipotent elements of G in relation to Iwasawa decompositions (see [Kos73, Propositions 2.3, 2.4, 2.5]) .

Theorem 2.16. (Kostant) *Let $G = KAN$ and $\mathfrak{g} = \mathfrak{k} \oplus \mathfrak{a} \oplus \mathfrak{n}$ be associated Iwasawa decompositions.*

(1) *An element $e \in G$ is elliptic if and only if it is conjugate to an element in K. Moreover, any element of $k \in K$ is of the form $k = \exp X$ for some $X \in \mathfrak{k}$.*

(2) *An element $X \in \mathfrak{g}$ is real semisimple if and only if it is conjugate to an element in $\mathfrak{a}$. Similarly, an element $h \in G$ is hyperbolic if and only if it is conjugated to an element in A.*

(3) *An element $u \in G$ is unipotent if and only if it is conjugate to an element in N.*

The following important result is called the *complete multiplicative Jordan decomposition*, abbreviated as CMJD (see [Kos73, Proposition 2.1]). It is an extension of Theorem 1.3 to Lie groups.

Theorem 2.17. (Kostant) *Each $g \in G$ can be uniquely written as*

$$g = ehu, \qquad (2.13)$$

where e is elliptic, h is hyperbolic, u is unipotent, and the three elements $e, h,$ and u commute.

Proof. Let $\mathfrak{g}$ be the (real) Lie algebra of G and let $\mathfrak{g}_\mathbb{C}$ be the complexification of $\mathfrak{g}$. Note that $\mathfrak{g}_\mathbb{C}$ is also semisimple as $\mathfrak{g}$ is semisimple [Hel78, p.132]. Let Int $\mathfrak{g}_\mathbb{C}$ be the adjoint group $\mathfrak{g}_\mathbb{C}$, i.e., the analytic subgroup of Aut $\mathfrak{g}_\mathbb{C}$ having ad $\mathfrak{g}_\mathbb{C}$ as its Lie algebra. Note that the adjoint representation Ad maps G into Int $\mathfrak{g}_\mathbb{C}$.

Let $g \in G$. Then $\operatorname{Ad} g$ is a Lie algebra homomorphism. In particular, it is a nonsingular linear transformation on $\mathfrak{g}_\mathbb{C}$ and we may use Theorem 1.3 to decompose it as

$$\operatorname{Ad} g = e'h'u' \qquad (2.14)$$

with e' elliptic, h' hyperbolic, and u' unipotent, with all three being unique and commuting.

Structure Theory of Semisimple Lie Groups

Note that both h' and u' define automorphisms on $\mathfrak{g}_\mathbb{C}$. Since $\mathfrak{g}_\mathbb{C}$ is semisimple, every derivation of $\mathfrak{g}_\mathbb{C}$ is $\operatorname{ad} Y$ for some $Y \in \mathfrak{g}_\mathbb{C}$ [Kna02, p.102]. Thus we have

$$h' = \exp \operatorname{ad} X \quad \text{and} \quad u' = \exp \operatorname{ad} Z \tag{2.15}$$

for some unique $X, Z \in \mathfrak{g}_\mathbb{C}$, where $\operatorname{ad} X$ is semisimple and $\operatorname{ad} Z$ is nilpotent. Now that h' and u' commute, so do $\operatorname{ad} X$ and $\operatorname{ad} Z$, which implies that

$$[X, Z] = 0. \tag{2.16}$$

Let $\sigma \in \operatorname{Int} \mathfrak{g}_\mathbb{C}$ be defined by

$$\sigma(A + iB) = A - iB, \qquad \forall A, B \in \mathfrak{g}.$$

Note that $\sigma^{-1} = \sigma$. For all $A, B \in \mathfrak{g}$, we have

$$\begin{aligned}
\sigma\left(\operatorname{Ad} g\right) \sigma(A+iB) &= \sigma \operatorname{Ad} g(A - iB) \\
&= \sigma(\operatorname{Ad}(g)A - i(\operatorname{Ad}(g)B) \\
&= \operatorname{Ad}(g)A + i(\operatorname{Ad}(g)B) \\
&= \operatorname{Ad} g(A + iB),
\end{aligned}$$

because $\mathfrak{g}$ is invariant under $\operatorname{Ad} g$ as a subspace of $\mathfrak{g}_\mathbb{C}$. Thus

$$\operatorname{Ad} g = \sigma \left(\operatorname{Ad} g\right) \sigma = \sigma e' h' u' \sigma = (\sigma e' \sigma)(\sigma h' \sigma)(\sigma u' \sigma).$$

Since $\sigma^{-1} = \sigma$, we have that $\sigma e' \sigma$, $\sigma h' \sigma$, and $\sigma u' \sigma$ remain elliptic, hyperbolic, and unipotent, respectively. Thus the uniqueness of the decomposition of $\operatorname{Ad} g$ in (2.14) yields

$$e' = \sigma e' \sigma, \quad h' = \sigma h' \sigma, \quad u' = \sigma u' \sigma.$$

Note that if $a \in \operatorname{Int} \mathfrak{g}_\mathbb{C}$ and if $\sigma a \sigma = a$, then $\mathfrak{g}$ is invariant under a:

$$a(Y) = \sigma a \sigma(Y) = \sigma a(Y), \qquad \forall Y \in \mathfrak{g}.$$

So $\mathfrak{g}$ is invariant under each of e', h', and u'. Consequently,

$$(\exp \operatorname{ad} X)Y = h'(Y) \in \mathfrak{g} \quad \text{and} \quad (\exp \operatorname{ad} Z)Y = u'(Y), \qquad \forall Y \in \mathfrak{g}.$$

Hence $\mathfrak{g}$ is invariant under $\operatorname{ad} X$ and $\operatorname{ad} Z$, implying that $X, Z \in \mathfrak{g}$.

Let $h = \exp X$ and $u = \exp Z$. Then h and u are hyperbolic and unipotent, respectively. Moreover, (2.16) implies that $hu = uh$ by the Baker-Campbell-Hausdorff formula [Hel78]. The commutativity of factors in the decomposition (2.14) of $\operatorname{Ad} g$ implies that $\operatorname{Ad} g$ commutes with h' and u'. Since $h' = \exp \operatorname{ad} X = \operatorname{Ad} h$ and $u' = \exp \operatorname{ad} Z = \operatorname{Ad} u$, we have $\operatorname{Ad} g$ commuting with $\operatorname{Ad} h$ and $\operatorname{Ad} u$. The kernel of Ad is the center of G [Hel78, p.129], so g commutes with h and u.

Let

$$e = g u^{-1} h^{-1}.$$

Then $g = ehu$ and e commutes with h and u. Moreover,

$$\operatorname{Ad} e = \operatorname{Ad} g (\operatorname{Ad} u)^{-1}(\operatorname{Ad} h)^{-1} = (\operatorname{Ad} g)(u')^{-1}(h')^{-1} = e',$$

implying that e is elliptic.

It remains to show the uniqueness of the decomposition (2.13) of g. Assume $g = e_1 h_1 u_1$ is another decomposition satisfying the same condition. By the uniqueness of the decomposition (2.14) of $\operatorname{Ad} g$, we have

$$\operatorname{Ad} e_1 = e', \quad \operatorname{Ad} h_1 = h', \quad \operatorname{Ad} u_1 = u'.$$

Now we have
$$\exp \operatorname{ad} \log h_1 = \operatorname{Ad} h_1 = h' = \exp \operatorname{ad} X,$$

so the uniqueness of X in (2.15) implies that $\log h_1 = X$. Similarly, $\log u_1 = Z$. Hence $h_1 = h$ and $u_1 = u$, which implies that $e_1 = e$. □

For each $g \in G$ with $g = ehu$ as in (2.13), we denote

$$e(g) = e, \quad h(g) = h, \quad u(g) = u$$

as the elliptic, hyperbolic, and unipotent components of g, respectively.

The following result is then obvious by the definition and the uniqueness of the CMJD.

Theorem 2.18. *Let $g \in G$ and let $g = ehu$ be the CMJD. Then the following statements are true.*

(1) $g^n = [e(g)]^n [h(g)]^n [u(g)]^n$ is the CMJD of g^n. In particular,

$$h(g^n) = [h(g)]^n.$$

(2) For any $f \in G$,

$$fgf^{-1} = [fe(g)f^{-1}][fh(g)f^{-1}][fu(g)f^{-1}]$$

is a CMJD. In particular, $h(fgf^{-1}) = fh(g)f^{-1}$. Furthermore, $f \in G$ commutes with g if and only if f commutes with each of e, h, and u.

(3) If $f \in G$ and $fg = gf$, then

$$e(fg) = e(f)e(g), \quad h(fg) = h(f)h(g), \quad u(fg) = u(f)u(g).$$

Let L denote the set of all hyperbolic elements in G. The following result describes L ([Kos73, Proposition 6.2]).

Theorem 2.19. (Kostant)

$$L = P^2 = \{pq : p, q \in P\}. \tag{2.17}$$

Proof. We first note that $P \subset L$, according to Theorem 2.16 and the fact that $\mathfrak{p} = \operatorname{Ad} K(\mathfrak{a})$.

To show that $P^2 \subset L$, we assume $p, q \in P$. Then by (2.6),

$$p^{1/2} q p^{1/2} = (p^{1/2} q^{1/2})(p^{1/2} q^{1/2})^* \in P,$$

hence it is conjugate to an element in A by (2.10). Since

$$pq = p^{1/2}(p^{1/2} q p^{1/2}) p^{-1/2},$$

it is also conjugate to an element in A, hence $pq \in L$ by Theorem 2.16.

To show that $L \subset P^2$, we assume $h \in L$. Then h is conjugate to some $d \in A \subset P$, say $h = gdg^{-1}$. Now

$$h = (gdg^*)(g^*)^{-1} g^{-1} = pq,$$

where $p = gdg^* = (gd^{1/2})(gd^{1/2})^* \in P$ and $q = (g^*)^{-1} g^{-1} = (gg^*)^{-1} \in P$ by (2.6). $\square$

Let $\mathfrak{l}$ denote the set of all real semisimple elements in $\mathfrak{g}$. The restriction of the exponential map on $\mathfrak{l}$ is then a bijection onto L. According to Theorem 2.16, $X \in \mathfrak{l}$ if and only if $\operatorname{Ad} g(X) \in \mathfrak{a}$ for some $g \in G$. Since $\mathfrak{p} = \operatorname{Ad} K(\mathfrak{a})$, we have

$$\mathfrak{l} = \operatorname{Ad} G(\mathfrak{a}) = \operatorname{Ad} G(\mathfrak{p}).$$

2.11 Kostant's Preorder

Let the notations be as in Section 2.10. Let G be a noncompact connected semisimple Lie group with Lie algebra $\mathfrak{g}$.

For any real semisimple element $X \in \mathfrak{g}$, let $W \cdot X$ denote the set of elements in $\mathfrak{a}$ that are conjugate to X, that is,

$$W \cdot X = \operatorname{Ad} G(X) \cap \mathfrak{a}. \tag{2.18}$$

It is known that $W \cdot X$ is a single W-orbit in $\mathfrak{a}$ ([Kos73, Proposition 2.4]). Let $\operatorname{conv} W \cdot X$ denote the convex hull in $\mathfrak{a}$ generated by $W \cdot X$. For any $g \in G$, denote

$$A(g) = \exp \operatorname{conv} W \cdot \log h(g), \tag{2.19}$$

where $h(g)$ is the hyperbolic component of g in its CMJD.

Kostant [Kos73, p.426] introduced the following preorder $\prec$ on G.

Definition 2.20. *Let $f, g \in G$. If $A(f) \subset A(g)$, then we say that*

$$f \prec g. \tag{2.20}$$

Kostant's preorder $\prec$ induces a partial order on the set of all conjugacy classes of G.

Kostant's preorder $\prec$ on G can be defined on $\mathfrak{g}$ as well, i.e., for $X, Y \in \mathfrak{g}$, define
$$X \prec Y \iff \exp X \prec \exp Y. \tag{2.21}$$
This preorder for $X, Y \in \mathfrak{l}$ means that
$$X \prec Y \iff \operatorname{conv} W \cdot X \subset \operatorname{conv} W \cdot Y. \tag{2.22}$$

Kostant's preorder does not depend on the choice of $\mathfrak{a}$ due to the following result ([Kos73, Theorem 3.1]).

Theorem 2.21. (Kostant) *Let $f, g \in G$. Then $f \prec g$ if and only if*
$$\rho(\pi(f)) \leqslant \rho(\pi(g)) \tag{2.23}$$
for all finite dimensional irreducible representations π of G, where $\rho(\pi(g))$ denotes the spectral radius of the operator $\pi(g)$.

Example 2.22. Consider the real simple Lie group $G = \operatorname{SL}_n(\mathbb{C})$. Let the Cartan decomposition $G = KP$ and the Iwasawa decomposition $G = KAN$ and the KA_+K decomposition be as in Example 2.7 and Example 2.12 and Example 2.15, respectively. As in Example 2.13, the Weyl group W is then isomorphic to S_n. The CMJD of G is illustrated in Theorem 1.3. In particular, if $g \in G$, then $h(g)$ is conjugate to
$$\operatorname{diag} |\lambda(g)| = \operatorname{diag} (|\lambda_1(g)|, \ldots, |\lambda_n(g)|) \in A_+,$$
where $\lambda_i(g)$'s are eigenvalues of g such that $|\lambda_1(g)| \geqslant \cdots \geqslant |\lambda_n(g)|$. This means that
$$A(g) = \exp \operatorname{conv} W \cdot \log h(g) = \exp \operatorname{conv} S_n \cdot \log(|\lambda(g)|).$$

Therefore, by Theorem 1.6, Kostant's preorder $\prec$ on G means that
$$f \prec g \iff |\lambda(f)| \prec_{\log} |\lambda(g)|.$$

The first statement of the following theorem is in [Kos73, p.448]. The second one can be shown in a similar manner.

Theorem 2.23. *The following statements are true for all $f, g \in G$.*

(1) If $f \prec g$, then $f^n \prec g^n$ for all $n \in \mathbb{N}$.

(2) If $f, g \in L$, then
$$f \prec g \iff f^r \prec g^r, \quad \forall r > 0. \tag{2.24}$$

Chapter 3

Inequalities for Matrix Exponentials

3.1 Golden-Thompson Inequality 67
3.2 Araki-Lieb-Thirring Inequality 75
3.3 Bernstein Inequality 76
3.4 Extensions to Lie Groups 80

With the Lie group decompositions introduced in the previous chapter, we now can extend matrix inequalities to connected noncompact real semisimple Lie groups. We start with inequalities involving matrix exponentials in this chapter.

3.1 Golden-Thompson Inequality

There are many inequalities involving the matrix exponential map. The Golden-Thompson inequality ([Gol65] and [Tho65]) is perhaps the most famous one among them.

Recall that $\mathbb{H}_n$ and $\mathbb{P}_n$ are the sets of all $n \times n$ Hermitian matrices and positive definite matrices, respectively, and that the exponential map

$$\exp : \mathbb{H}_n \to \mathbb{P}_n$$

is bijective.

Theorem 3.1. (Golden-Thompson) *For all $A, B \in \mathbb{H}_n$*

$$\operatorname{tr} e^{A+B} \leqslant \operatorname{tr} e^A e^B. \tag{3.1}$$

Moreover, equality holds if and only if $AB = BA$.

To give a simple proof of Theorem 3.1, we need the following two lemmas.

Lemma 3.2. *For all $X \in \mathbb{C}_{n \times n}$ and $m \in \mathbb{N}$, we have*

$$|\operatorname{tr} X^m| \leqslant \operatorname{tr} |X|^m,$$

where $|X| = (X^*X)^{1/2}$.

Proof. If $X \in \mathbb{C}_{n \times n}$ and $m \in \mathbb{N}$, then

$$|\operatorname{tr} X^m| = \left|\sum_{i=1}^{n} [\lambda_i(X)]^m\right| \leqslant \sum_{i=1}^{n} |\lambda_i(X)|^m \leqslant \sum_{i=1}^{n} [s_i(X)]^m = \operatorname{tr} |X|^m,$$

because $(|\lambda_1(X)|^p, \ldots, |\lambda_n(X)|^p) \prec_w (s_1(X)^p, \ldots, s_n(X)^p)$ for all $p > 0$, according to Weyl's Majorant Theorem (see [Bha97, p.42]). $\square$

Lemma 3.3. *For any $P, Q \in \mathbb{P}_n$ and $k \in \mathbb{N}$, we have*

$$\operatorname{tr}(PQ)^{2^k} \leqslant \operatorname{tr} P^{2^k} Q^{2^k},$$

and the function $k \mapsto \operatorname{tr}(P^{1/2^k} Q^{1/2^k})^{2^k}$ is monotonically decreasing as $k \to \infty$.

Proof. Applying Lemma 3.2 on $X = PQ$ and $m = 2^k$, we have

$$\operatorname{tr}(PQ)^{2^k} \leqslant \operatorname{tr}(PQ^2 P)^{2^{k-1}} = \operatorname{tr}(P^2 Q^2)^{2^{k-1}}.$$

Applying Lemma 3.2 again on $X = P^2 Q^2$ and $m = 2^{k-1}$, we have

$$\operatorname{tr}(P^2 Q^2)^{2^{k-1}} \leqslant \operatorname{tr}(P^2 Q^4 P^2)^{2^{k-2}} = \operatorname{tr}(P^4 Q^4)^{2^{k-2}}.$$

Continuing the process, we see

$$\operatorname{tr}(PQ)^{2^k} \leqslant \operatorname{tr}(P^2 Q^2)^{2^{k-1}} \leqslant \cdots \leqslant \operatorname{tr}(P^{2^k} Q^{2^k}). \tag{3.2}$$

Now by replacing P and Q with $P^{1/2^k}$ and $Q^{1/2^k}$, respectively, in (3.2), we conclude that the function $k \mapsto \operatorname{tr}(P^{1/2^k} Q^{1/2^k})^{2^k}$ is monotonically decreasing as $k \to \infty$. $\square$

Proof of Theorem 3.1. Applying Lemma 3.3 on $P = e^{A/2^k}$ and $Q = e^{B/2^k}$ for $k \in \mathbb{N}$, we obtain that

$$\operatorname{tr}(e^{A/2^k} e^{B/2^k})^{2^k} \leqslant \operatorname{tr}(e^{A/2^k})^{2^k} (e^{B/2^k})^{2^k} = \operatorname{tr} e^A e^B.$$

Letting $k \to \infty$ yields Theorem 3.1 by the Lie product formula.

For the nontrivial equality case, suppose $\operatorname{tr} e^{A+B} = \operatorname{tr} e^A e^B$. By Lemma 3.3, $\operatorname{tr}(e^{A/2^k} e^{B/2^k})^{2^k}$ decreases monotonically to $\operatorname{tr} e^{A+B}$. In particular, we have $\operatorname{tr} e^A e^B = \operatorname{tr}(e^{A/2} e^{B/2})^2$, which implies by direct computation that

$$\operatorname{tr}(e^{A/2} e^{B/2} - e^{B/2} e^{A/2})(e^{A/2} e^{B/2} - e^{B/2} e^{A/2})^* = 0.$$

Thus $e^{A/2} e^{B/2} = e^{B/2} e^{A/2}$. By Theorem 1.12 (8), we conclude that $AB = BA$. $\square$

Motivated by the equality case of the Golden-Thompson inequality, we state the following result.

Corollary 3.4. *The following statements are equivalent for $A, B \in \mathbb{H}_n$.*

(1) $AB = BA$.

(2) $e^{A+B} = e^A e^B$.

(3) $\lambda(e^{A+B}) = \lambda(e^A e^B)$.

(4) $\operatorname{tr} e^{A+B} = \operatorname{tr} e^A e^B$.

(5) $e^A e^B = e^B e^A$.

Proof. The implications (1) $\Rightarrow$ (2) $\Rightarrow$ (3) $\Rightarrow$ (4) are obvious, and (4) $\Rightarrow$ (1) is part of Theorem 3.1. The equivalence (1) $\Leftrightarrow$ (5) is Theorem 1.12 (8). □

For general $A, B \in \mathbb{H}_n$, while $e^A e^B$ is not positive definite, $e^{A/2} e^B e^{A/2}$ is. Note that $e^A e^B$ and $e^{A/2} e^B e^{A/2}$ have the same eigenvalues, because $\lambda(XY) = \lambda(YX)$ for all $X, Y \in \mathbb{C}_{n \times n}$. Thus the Golden-Thompson inequality (3.1) is equivalent to

$$\operatorname{tr} e^{A+B} \leqslant \operatorname{tr} e^{A/2} e^B e^{A/2}, \qquad \forall\, A, B \in \mathbb{H}_n. \tag{3.3}$$

The following result is stronger than the Golden-Thompson inequality.

Theorem 3.5. *Let $A, B \in \mathbb{H}_n$. The following statements are true and equivalent.*

(1) $\lambda(e^{A+B}) \prec_{\log} \lambda(e^{A/2} e^B e^{A/2}) = \lambda(e^A e^B)$.

(2) $\lambda(e^{A+B}) \prec_w \lambda(e^{A/2} e^B e^{A/2}) = \lambda(e^A e^B)$.

(3) $s(e^{A+B}) \prec_w s(e^{A/2} e^B e^{A/2}) \prec_w s(e^A e^B)$.

(4) $|||e^{A+B}||| \leqslant |||e^{A/2} e^B e^{A/2}||| \leqslant |||e^A e^B|||$ *for any unitarily invariant norm $||| \cdot |||$ on $\mathbb{C}_{n \times n}$.*

(5) $\|e^{A+B}\| \leqslant \|e^{A/2} e^B e^{A/2}\| \leqslant \|e^A e^B\|$, *where $\| \cdot \|$ is the spectral norm on $\mathbb{C}_{n \times n}$.*

(6) $\lambda_1(e^{A+B}) \leqslant \lambda_1(e^{A/2} e^B e^{A/2}) = \lambda_1(e^A e^B)$.

In particular, the Golden-Thompson inequality (3.1) follows from (2) and hence from each of (1)–(6).

Proof. The implications (1) $\Rightarrow$ (2) $\Rightarrow$ (3) $\Rightarrow$ (4) $\Rightarrow$ (5) $\Rightarrow$ (6) are obvious, since

$$\lambda(e^{A+B}) = s(e^{A+B}) \quad \text{and} \quad \lambda(e^{A/2} e^B e^{A/2}) = s(e^{A/2} e^B e^{A/2}).$$

In (2) $\Rightarrow$ (3), Theorem 1.17 and Theorem 1.7 are used to show that $\lambda(e^A e^B) \prec_w s(e^A e^B)$. And (3) $\Rightarrow$ (4) follows from the Fan Dominance Theorem (Theorem 1.11). It remains to show that (6) $\Rightarrow$ (1) and that (6) is valid.

To show (6) ⇒ (1), we apply compound matrix arguments. Suppose
$$\lambda_1(e^{A+B}) \leqslant \lambda_1(e^A e^B), \qquad \forall A, B \in \mathbb{H}_n.$$
For each $k \in \mathbb{N}$, by Theorem 1.15, (1.33), and Theorem 1.16, we have

$$\prod_{i=1}^{k} \lambda_i(e^{A+B}) = \lambda_1(C_k(e^{A+B}))$$
$$= \lambda_1(e^{\Delta_k(A+B)})$$
$$= \lambda_1(e^{\Delta_k(A)+\Delta_k(B)})$$
$$\leqslant \lambda_1(e^{\Delta_k(A)} e^{\Delta_k(B)})$$
$$= \lambda_1(C_k(e^A) C_k(e^B))$$
$$= \lambda_1(C_k(e^A e^B))$$
$$= \prod_{i=1}^{k} \lambda_i(e^A e^B).$$

In other words, $\lambda(e^{A+B}) \prec_{\text{w-log}} \lambda(e^A e^B)$. Note also that
$$\det(e^{A+B}) = e^{\operatorname{tr}(A+B)} = e^{\operatorname{tr} A} e^{\operatorname{tr} B} = \det(e^A)\det(e^B) = \det(e^A e^B).$$
Therefore, $\lambda(e^{A+B}) \prec_{\log} \lambda(e^A e^B)$.
The fact that (6) is valid follows from Theorem 3.7 below. □

In Theorem 3.5 (4), if we choose the unitarily invariant norm $|||\cdot|||$ to be the Schatten p–norm, defined as
$$\|X\|_p = \left(\sum_{i=1}^{n} [s_i(X)]^p \right)^{1/p}, \qquad \forall X \in \mathbb{C}_{n \times n},$$
then (4) becomes
$$\|e^{A+B}\|_p \leqslant \|e^{A/2} e^B e^{A/2}\|_p \leqslant \|e^A e^B\|_p,$$
where the first inequality is equivalent to (replacing A with A/p)
$$\operatorname{tr} e^{A+B} \leqslant \operatorname{tr}(e^{A/2p} e^{B/p} e^{A/2p})^p = \operatorname{tr}(e^{A/p} e^{B/p})^p, \qquad \forall p \geqslant 1. \qquad (3.4)$$
In particular, the Golden-Thompson inequality follows from (3.4) when $p = 1$.

Theorem 3.5 in the form of (1) can be generalized to normal matrices (see Theorem 3.17) and it also has an extension in Lie groups (see Theorem 3.25).

Recall that $\|X\| = \|XX^*\|^{1/2}$ for all $X \in \mathbb{C}_{n \times n}$, where $\|\cdot\|$ denotes the spectral norm. Therefore,
$$\|(AB)\|^2 = \|AB^2 A\| = \lambda_1(AB^2 A) = \lambda_1(A^2 B^2), \qquad \forall A, B \in \mathbb{H}_n. \qquad (3.5)$$

The following result is of fundamental importance ([Cor87]).

Theorem 3.6. (Cordes) *Let $A, B \in \mathbb{P}_n$ and let $\|\cdot\|$ denote the spectral norm. The following statements are true and equivalent.*

(1) $\|A^r B^r\| \leqslant \|AB\|^r$ for all $0 \leqslant r \leqslant 1$.

(2) $\|A^r B^r\| \geqslant \|AB\|^r$ for all $r \geqslant 1$.

(3) The function $r \mapsto \|A^r B^r\|^{1/r}$ is monotonically increasing on $(0, \infty)$.

(4) The function $r \mapsto \|A^{1/r} B^{1/r}\|^r$ is monotonically decreasing on $(0, \infty)$.

Proof. We first show that (1) is valid. Let

$$S = \{r \in [0,1] : \|A^r B^r\| \leqslant \|AB\|^r\}.$$

Obviously, $0 \in S$ and $1 \in S$. We will see that $S = [0,1]$.

If $\|A^r B^r\| \leqslant \|AB\|^r$ and $\|A^t B^t\| \leqslant \|AB\|^t$ for some $0 \leqslant r \leqslant t \leqslant 1$, then

$$\begin{aligned}
\|A^{(r+t)/2} B^{(r+t)/2}\|^2 &= \lambda_1(A^{r+t} B^{r+t}) \quad \text{(by (3.5))} \\
&= \lambda_1(A^r B^{r+t} A^t) \\
&\leqslant \|A^r B^{r+t} A^t\| \\
&\leqslant \|A^r B^r\| \cdot \|B^t A^t\| \\
&= \|A^r B^r\| \cdot \|A^t B^t\| \quad (\|X\| = \|X^*\| \text{ for all } X \in \mathbb{C}_{n \times n}) \\
&\leqslant \|AB\|^r \cdot \|AB\|^t \\
&= \|AB\|^{r+t},
\end{aligned}$$

and thus $\|A^{(r+t)/2} B^{(r+t)/2}\| \leqslant \|AB\|^{(r+t)/2}$. This shows that S is a convex set. Therefore, $S = [0,1]$ and hence (1) is valid.

(1) $\Rightarrow$ (2). By (1), we have $\|A^{1/r} B^{1/r}\| \leqslant \|AB\|^{1/r}$ for all $r \geqslant 1$. So $\|A^{1/r} B^{1/r}\|^r \leqslant \|AB\|$ for all $r \geqslant 1$. Replacing A with A^r and B with B^r, respectively, we derive (2).

(2) $\Rightarrow$ (3). By (2), we have $\|A^{p/q} B^{p/q}\| \geqslant \|AB\|^{p/q}$ for all $p \geqslant q > 0$. So $\|A^{p/q} B^{p/q}\|^{1/p} \geqslant \|AB\|^{1/q}$ for all $p \geqslant q > 0$. Replacing A with A^q and B with B^q, respectively, we derive (3).

(3) $\Rightarrow$ (4). This is obvious.

(4) $\Rightarrow$ (1). By (4), we have $\|A^{1/r} B^{1/r}\|^r \geqslant \|AB\|$ for all $0 < r < 1$. Replacing A with A^r and B with B^r, respectively, we derive (1). $\square$

Although $\|AB\| \neq \lambda_1(AB)$ for general $A, B \in \mathbb{P}_n$, the following result is equivalent to Theorem 3.6.

Theorem 3.7. *Let $A, B \in \mathbb{P}_n$. The following statements are true and equivalent.*

(1) $\lambda_1(A^r B^r) \leqslant \lambda_1((AB)^r)$ for all $0 \leqslant r \leqslant 1$.

(2) $\lambda_1(A^r B^r) \geqslant \lambda_1((AB)^r)$ for all $r \geqslant 1$.

(3) The function $r \mapsto \lambda_1((A^r B^r)^{1/r})$ is monotonically increasing on $(0, \infty)$.

(4) The function $r \mapsto \lambda_1((A^{1/r} B^{1/r})^r)$ is monotonically decreasing on $(0, \infty)$.

Proof. The proof of the equivalence of (1)–(4) is similar to that of Theorem 3.6. We only show that (1) is true. By (3.5) and Theorem 3.6, we have that for all $0 \leqslant r \leqslant 1$,

$$\lambda_1(A^{2r} B^{2r}) = \|A^r B^r\|^2 \leqslant \|AB\|^{2r} = [\lambda_1(A^2 B^2)]^r = \lambda_1((A^2 B^2)^r).$$

Replacing A^2 with A and B^2 with B, respectively, we derive (1). □

Proof of Theorem 3.5 (6). By Theorem 3.7 and the Lie product formula, we have

$$\lambda_1(e^{A+B}) = \lambda_1\left(\lim_{n\to\infty} (e^{A/n} e^{B/n})^n\right) = \lim_{n\to\infty} \lambda_1((e^{A/n} e^{B/n})^n) \leqslant \lambda_1(e^A e^B).$$

This completes the proof of Theorem 3.5. □

Using arguments involving compound matrices, we obtain the following stronger results.

Theorem 3.8. *Let $A, B \in \mathbb{P}_n$. The following statements are true and equivalent.*

(1) $\lambda(A^r B^r) \prec_{\log} \lambda((AB)^r)$ for all $0 \leqslant r \leqslant 1$.

(2) $\lambda((AB)^r) \prec_{\log} \lambda(A^r B^r)$ for all $r \geqslant 1$.

(3) The function $r \mapsto \lambda((A^r B^r)^{1/r})$ is monotonically increasing on $(0, \infty)$ in terms of $\prec_{\log}$. In other words,

$$\lambda((A^r B^r)^{1/r}) \prec_{\log} \lambda((A^t B^t)^{1/t}), \qquad \forall\, 0 < r < t.$$

Moreover, the function $r \mapsto \lambda((A^r B^r)^{1/r})$ is not bounded above.

(4) The function $r \mapsto \lambda((A^{1/r} B^{1/r})^r)$ is monotonically decreasing on $(0, \infty)$ in terms of $\prec_{\log}$. Moreover, the limit is $\lambda(e^{\log A + \log B})$.

Proof. The proof of the equivalence of (1)–(4) is similar to that of Theorem 3.6. The unboundedness of the function in (3) follows from the Baker-Campbell-Hausdorff formula. The limit in (4) is guaranteed by the Lie product formula. We only need to show that (1) is valid.

Let $0 \leqslant r \leqslant 1$. For each $1 \leqslant k \leqslant n$, by Theorems 1.15, we have

$$\begin{aligned}
\prod_{i=1}^{k} \lambda_i(A^r B^r) &= \lambda_1(C_k(A^r B^r)) \\
&= \lambda_1(C_k(A^r) C_k(B^r)) \\
&= \lambda_1([C_k(A)]^r [C_k(B)]^r) \\
&\leqslant \lambda_1([C_k(A) C_k(B)]^r) \qquad \text{(by Theorem 3.7)} \\
&= \lambda_1([C_k(AB)]^r) \\
&= \lambda_1(C_k((AB)^r)) \\
&= \prod_{i=1}^{k} \lambda_i((AB)^r).
\end{aligned}$$

In other words, $\lambda(A^r B^r) \prec_{\text{w-log}} \lambda((AB)^r)$. Note also that

$$\det A^r B^r = \det (AB)^r.$$

Therefore, $\lambda(A^r B^r) \prec_{\log} \lambda((AB)^r)$. This completes the proof of (1). □

The following result combines Theorem 3.7 and Theorem 3.8, in a pattern similar to that of Theorem 3.5.

Theorem 3.9. *Let $A, B \in \mathbb{P}_n$. The following statements are true and equivalent.*

(1) The function $r \mapsto [\lambda(A^r B^r)]^{1/r}$ is monotonically increasing on $(0, \infty)$ in terms of $\prec_{\log}$.

(2) The function $r \mapsto [\lambda(A^r B^r)]^{1/r}$ is monotonically increasing on $(0, \infty)$ in terms of $\prec_w$.

(3) The function $r \mapsto [\lambda_1(A^r B^r)]^{1/r}$ is monotonically increasing on $(0, \infty)$.

(4) The function $r \mapsto [s(A^r B^r)]^{1/r}$ is monotonically increasing on $(0, \infty)$ in terms of $\prec_{\log}$.

(5) The function $r \mapsto |||A^r B^r|||^{1/r}$ is monotonically increasing on $(0, \infty)$ for any unitarily invariant norm $||| \cdot |||$.

(6) The function $r \mapsto \|A^r B^r\|^{1/r}$ is monotonically increasing on $(0, \infty)$ for the spectral norm $\| \cdot \|$.

(7) The function $r \mapsto [s_1(A^r B^r)]^{1/r}$ is monotonically increasing on $(0, \infty)$.

Proof. The implications $(1) \Rightarrow (2) \Rightarrow (3) \Rightarrow (1)$ follow by the equivalence of Theorem 3.7 and Theorem 3.8. The implications $(4) \Rightarrow (5) \Rightarrow (6) \Rightarrow (7)$ are obvious. The equivalence $(3) \Leftrightarrow (6)$ follows from (3.5), as partially shown in the proof of Theorem 3.7 (1). It remains to show that $(7) \Rightarrow (4)$.

Let $0 < r < t$. Then (7) means that
$$[s_1(A^r B^r)]^{1/r} \leqslant [s_1(A^t B^t)]^{1/t}. \tag{3.6}$$

For each $1 \leqslant k \leqslant n$, by Theorems 1.15, we have
$$\begin{aligned}
\prod_{i=1}^{k}[s_i(A^r B^r)]^{1/r} &= \left(\prod_{i=1}^{k} s_i(A^r B^r)\right)^{1/r} \\
&= [s_1(C_k(A^r B^r))]^{1/r} \\
&= [s_1([C_k(A)]^r [C_k(B)]^r)]^{1/r} \\
&\leqslant [s_1([C_k(A)]^t [C_k(B)]^t)]^{1/t} \quad \text{(by (3.6))} \\
&= \left(\prod_{i=1}^{k} s_i(A^t B^t)\right)^{1/t} \\
&= \prod_{i=1}^{k}[s_i(A^t B^t)]^{1/t}.
\end{aligned}$$

In other words, $[s(A^r B^r)]^{1/r} \prec_{\text{w-log}} [s(A^t B^t)]^{1/t}$. Note also that
$$\prod_{i=1}^{n}[s_i(A^r B^r)]^{1/r} = [\det(A^r B^r)]^{1/r} = \det(A)\det(B) = \prod_{i=1}^{n}[s_i(A^t B^t)]^{1/t}.$$

Therefore, $[s(A^r B^r)]^{1/r} \prec_{\log} [s(A^t B^t)]^{1/t}$. This shows that $(7) \Rightarrow (4)$. □

Although Theorem 3.9 is about $(A^r B^r)^{1/r}$, one may also formulate similar results for $(A^{1/r} B^{1/r})^r$ and $A^r B^r$ and $(AB)^r$, based on Theorem 3.7 (or Theorem 3.8).

If $A, B \in \mathbb{H}_n$, then $\operatorname{tr}(e^{A/2^k} e^{B/2^k})^{2^k}$ decreases monotonically to $\operatorname{tr} e^{A+B}$ as $k \to \infty$ by Lemma 3.3. According to Theorem 3.9, the term 2^k can be replaced by any positive number. The following result is then obvious, from which the Golden-Thompson inequality follows.

Corollary 3.10. *Let $A, B \in \mathbb{P}_n$. The following statements are true and equivalent.*

(1) $\operatorname{tr} A^r B^r \leqslant \operatorname{tr}(AB)^r$ for all $0 \leqslant r \leqslant 1$.

(2) $\operatorname{tr} A^r B^r \geqslant \operatorname{tr}(AB)^r$ for all $r \geqslant 1$.

(3) The function $r \mapsto \operatorname{tr}(A^r B^r)^{1/r}$ is monotonically increasing without bound on $(0, \infty)$.

(4) The function $r \mapsto \operatorname{tr}(A^{1/r} B^{1/r})^r$ is monotonically decreasing on $(0, \infty)$, and its limit is $\operatorname{tr} e^{\log A + \log B}$.

Notes and References: The celebrated Golden-Thompson inequality was independently discovered by Golden [Gol65], Symanzik [Sym65], and Thompson [Tho65] in the same year of 1965, all motivated by statistical mechanics. Since then, the Golden-Thompson inequality has received intensive attention, and has been generalized in various ways and applied in many fields (see, for instance, [AH94, Ber88, Bha97, CFKK82, FT14, Hia97, HP93, Kos73, Len71, Pet94, Tho71] and the references therein). For historical aspects, one may see a recent paper by Forrester and Thompson [FT14]. The case for equality was mentioned in [Len71] and established in [So92].

3.2 Araki-Lieb-Thirring Inequality

Another famous trace inequality having many applications is the Araki-Lieb-Thirring inequality ([Ara90] and [LT76]).

Theorem 3.11. (Araki-Lieb-Thirring) *Suppose $A, B \in \mathbb{C}_{n \times n}$ are positive semidefinite. Then*

$$\operatorname{tr}(A^{1/2}BA^{1/2})^{rq} \leqslant \operatorname{tr}(A^{r/2}B^r A^{r/2})^q, \qquad \forall q \geqslant 0, \forall r \geqslant 1, \tag{3.7}$$

$$\operatorname{tr}(A^{1/2}BA^{1/2})^{rq} \geqslant \operatorname{tr}(A^{r/2}B^r A^{r/2})^q, \qquad \forall q \geqslant 0, \forall 0 \leqslant r \leqslant 1. \tag{3.8}$$

Proof. Without loss of generality, we may assume that $A, B \in \mathbb{P}_n$, for the case of positive semidefinite matrices can be shown by a continuity argument. Moreover, we only need to show the case of $r \geqslant 1$, for the other case is similar.

Note that $\lambda((A^{1/2}BA^{1/2})^r) = \lambda((AB)^r)$ and $\lambda(A^{r/2}B^r A^{r/2}) = \lambda(A^r B^r)$. Therefore, Theorem 3.8 is equivalent to

$$\lambda((A^{1/2}BA^{1/2})^r) \prec_{\log} \lambda(A^{r/2}B^r A^{r/2}), \qquad \forall r \geqslant 1. \tag{3.9}$$

Thus we have

$$\lambda((A^{1/2}BA^{1/2})^{rq}) \prec_{\log} \lambda((A^{r/2}B^r A^{r/2})^q), \qquad \forall q \geqslant 0, \forall r \geqslant 1.$$

It follows that

$$\lambda((A^{1/2}BA^{1/2})^{rq}) \prec_w \lambda((A^{r/2}B^r A^{r/2})^q), \qquad \forall q \geqslant 0, \forall r \geqslant 1.$$

Hence (3.7) is valid. □

It is worthwhile to formulate some equivalent forms of (3.9), which are corresponding and equivalent to those in Theorem 3.8.

Theorem 3.12. *Let $A, B \in \mathbb{P}_n$. The following statements are true and equivalent.*

(1) $\lambda(A^{r/2}B^r A^{r/2}) \prec_{\log} \lambda((A^{1/2}BA^{1/2})^r)$ for all $0 \leq r \leq 1$.

(2) $\lambda((A^{1/2}BA^{1/2})^r) \prec_{\log} \lambda(A^{r/2}B^r A^{r/2})$ for all $r \geq 1$.

(3) The function $r \mapsto \lambda((A^{r/2}B^r A^{r/2})^{1/r})$ is monotonically increasing on $(0, \infty)$ in terms of $\prec_{\log}$.

(4) The function $r \mapsto \lambda((A^{1/2r}B^{1/r}A^{1/2r})^r)$ is monotonically decreasing on $(0, \infty)$ in terms of $\prec_{\log}$. Moreover, the limit is $\lambda(e^{\log A + \log B})$.

Remark 3.13. The statements in Theorems 3.6, 3.7, 3.8, 3.9, and 3.12 are all equivalent to each other. Combined with the Lie product formula, each of them implies Theorem 3.5. Because of the bijection $X \mapsto e^X$ from $\mathbb{H}_n$ onto $\mathbb{P}_n$, they can be expressed with a form involving the matrix exponential map. Furthermore, in the form of $\prec_{\log}$, they have extensions in Lie groups.

Notes and References: The following Lieb-Thirring inequality was first established in [LT76]:

$$\operatorname{tr}(A^{1/2}BA^{1/2})^r \leq \operatorname{tr}(A^{r/2}B^r A^{r/2}), \qquad \forall r \geq 1.$$

Then Araki proved a much more general result in [Ara90], of which Theorem 3.11 is a special case.

3.3 Bernstein Inequality

Motivated by the Golden-Thompson inequality (3.1) and problems in linear-quadratic optimal feedback control, Bernstein proved the following trace inequality ([Ber88]).

Theorem 3.14. (Bernstein) *For all $A \in \mathbb{C}_{n \times n}$*

$$\operatorname{tr} e^{A^*}e^A \leq \operatorname{tr} e^{A^*+A}. \tag{3.10}$$

Equality holds if and only if $A \in \mathbb{N}_n$.

The Bernstein inequality (3.10) follows from (3.15) in Theorem 3.16 below. The equality case in Theorem 3.14 was shown in [So92] (recall that $\mathbb{N}_n$ denotes the set of all normal matrices in $\mathbb{C}_{n \times n}$).

Recall that $|A| = (A^*A)^{1/2}$ for all $A \in \mathbb{C}_{n \times n}$. The following result is important ([Fan49]).

Inequalities for Matrix Exponentials

Theorem 3.15. (Fan) *Let $A \in \mathbb{C}_{n \times n}$. Then the following two relations are equivalent and valid for all $m \in \mathbb{N}$:*

$$\lambda((A^m)^* A^m) \prec_{\log} \lambda((A^* A)^m), \tag{3.11}$$

$$\lambda(|A^m|) = s(A^m) \prec_{\log} [s(A)]^m = [\lambda(|A|)]^m. \tag{3.12}$$

Consequently,

$$\operatorname{tr}(A^m)^* A^m \leqslant \operatorname{tr}(A^* A)^m, \quad \forall\, m \in \mathbb{N}. \tag{3.13}$$

Moreover, $\operatorname{tr}(A^m)^* A^m = \operatorname{tr}(A^* A)^m$ for some $m \geqslant 2$ if and only if $A \in \mathbb{N}_n$.

Proof. The equivalence of (3.11) and (3.12) follows immediately, noting that $s(X) = \lambda(|X|)$ for all $X \in \mathbb{C}_{n \times n}$. It is known [Fan49, Theorem 3] (see [Coh88, Theorem 1] for some other interesting inequalities) that

$$\sum_{i=1}^{k} \lambda_i((A^m)^* A^m) \leqslant \sum_{i=1}^{k} \lambda_i((A^* A)^m), \quad 1 \leqslant k \leqslant n.$$

In other words, $\lambda((A^m)^* A^m) \prec_w \lambda((A^* A)^m)$. In particular, we have

$$\lambda_1((A^m)^* A^m) \leqslant \lambda_1((A^* A)^m). \tag{3.14}$$

Then by a compound matrix argument, we see that

$$\prod_{i=1}^{k} \lambda_i((A^m)^* A^m) = \lambda_1(C_k((A^m)^* A^m))$$

$$= \lambda_1(C_k((A^m)^*) C_k(A^m))$$

$$= \lambda_1(([C_k(A)]^*)^m [C_k(A)]^m)$$

$$\leqslant \lambda_1(([C_k(A)]^* [C_k(A)])^m) \quad \text{(by (3.14))}$$

$$= \lambda_1([C_k(A^* A)]^m)$$

$$= \lambda_1(C_k((A^* A)^m))$$

$$= \prod_{i=1}^{k} \lambda_i((A^* A)^m).$$

Obviously, $\det(A^m)^* A^m = \det(A^* A)^m$. Thus (3.11) is established.

If $A \in \mathbb{N}_n$, then $(A^m)^* A^m = (A^*)^m A^m = (A^* A)^m$ for all $m \in \mathbb{N}$, hence $\operatorname{tr}(A^m)^* A^m = \operatorname{tr}(A^* A)^m$ for all $m \in \mathbb{N}$. The converse is [So92, Theorem 4.4]. □

Theorem 3.15 can be extended to Lie groups (see Theorem 3.33).

As an application of Theorem 3.15, the following result in [Coh88, Theorem 2] is a generalization of the Bernstein inequality (3.10). It is a matrix version of the scalar identity $|e^{x+iy}| = e^x$ for $x, y \in \mathbb{R}$.

Theorem 3.16. (Cohen) *For all $A \in \mathbb{C}_{n \times n}$, the following two relations are equivalent and valid:*

$$\lambda(e^{A^*} e^A) \prec_{\log} \lambda(e^{A^*+A}), \qquad (3.15)$$

$$\lambda(|e^A|) = s(e^A) \prec_{\log} s(e^{\operatorname{Re} A}) = \lambda(e^{\operatorname{Re} A}), \qquad (3.16)$$

where $\operatorname{Re} A = (A^* + A)/2$ is the Hermitian part of A.

Proof. Obviously, (3.15) and (3.16) are equivalent. Applying (3.11) on $e^{A/m}$ and noting that $(e^A)^* = e^{A^*}$, we get

$$\lambda(e^{A^*} e^A) \prec_{\log} \lambda([e^{A^*/m} e^{A/m}]^m), \qquad \forall m \in \mathbb{N}. \qquad (3.17)$$

Then (3.15) is established by the Lie product formula. $\square$

Theorem 3.16 has an extension in Lie groups (see Theorem 3.34).

The following result is a generalization of Theorem 3.5 to normal matrices.

Theorem 3.17. *Let $A, B \in \mathbb{N}_n$. Then*

$$\lambda(|e^{A+B}|) \prec_{\log} \lambda(|e^A| \cdot |e^B|). \qquad (3.18)$$

Proof. Applying Theorem 3.16 on $A + B$, we have

$$\lambda(e^{(A+B)^*} e^{A+B}) \prec_{\log} \lambda(e^{(A^*+A)+(B^*+B)}). \qquad (3.19)$$

By Theorem 3.5 and the normality of A and B, we have

$$\lambda(e^{(A^*+A)+(B^*+B)}) \prec_{\log} \lambda(e^{A^*+A} e^{B^*+B}) = \lambda(e^{A^*} e^A e^{B^*} e^B). \qquad (3.20)$$

Combining (3.19) and (3.20), we derive (3.18), because eigenvalues respect product and power. $\square$

Similarly, Theorem 3.12 can be generalized to normal matrices.

Theorem 3.18. *Let $A, B \in \mathbb{N}_n$. Then*

$$\lambda(|e^{rA/2}| \cdot |e^{rB}| \cdot |e^{rA/2}|) \prec_{\log} \lambda((|e^{A/2}| \cdot |e^B| \cdot |e^{A/2}|)^r), \qquad \forall\, 0 \leqslant r \leqslant 1, \qquad (3.21)$$

$$\lambda((|e^{A/2}| \cdot |e^B| \cdot |e^{A/2}|)^r) \prec_{\log} \lambda(|e^{rA/2}| \cdot |e^{rB}| \cdot |e^{rA/2}|), \qquad \forall\, r \geqslant 1. \qquad (3.22)$$

Proof. Note that $|e^A| = e^{(A^*+A)/2}$ by the normality of A, with $(A^* + A)/2 \in \mathbb{H}_n$. Therefore, (3.21) and (3.22) follow from Theorem 3.12 (1) and (2), respectively. $\square$

For each $A \in \mathbb{C}_{n \times n}$, there are two naturally associated Hermitian matrices, namely, $A^* + A$ and $A^* A$. The following inequality is interesting ([FH55, p.114]).

Theorem 3.19. (Fan) *For any $A \in \mathbb{C}_{n \times n}$, we have*

$$\lambda_k \left(\frac{A^* + A}{2} \right) \leqslant \lambda_k \left((A^*A)^{1/2} \right), \qquad \forall \, 1 \leqslant k \leqslant n.$$

In other words, $\lambda_k(\operatorname{Re} A) \leqslant s_k(A)$ for all $1 \leqslant k \leqslant n$.

Proof. Let x_k and y_k be nonzero vectors in $\mathbb{C}^n$ such that for all $1 \leqslant k \leqslant n$

$$(\operatorname{Re} A)x_k = \lambda_k(\operatorname{Re} A)x_k \quad \text{and} \quad (A^*A)^{1/2} y_k = s_k(A) y_k.$$

Because $\operatorname{Re} A$ is Hermitian, $x_1, \ldots, x_n$ can be chosen so that they are orthogonal with respect to the standard inner product $\langle \cdot, \cdot \rangle$, even if the $\lambda_k(\operatorname{Re} A)$'s are not distinct. Similarly, we chose $y_1, \ldots, y_n$ to be linearly independent. For each k, since

$$\dim \operatorname{span}\{x_1, \ldots, x_k\} + \dim \operatorname{span}\{y_k, \ldots, y_n\} = n + 1,$$

we have $\dim(\operatorname{span}\{x_1, \ldots, x_k\} \cap \operatorname{span}\{y_k, \ldots, y_n\}) \geqslant 1$.

Pick $z \in \operatorname{span}\{x_1, \ldots, x_k\} \cap \operatorname{span}\{y_k, \ldots, y_n\}$ with $\|z\|_2 = 1$. Then we have

$$\lambda_k \leqslant \langle z, (\operatorname{Re} A)z \rangle = \operatorname{Re} \langle z, Az \rangle \leqslant |\langle z, Az \rangle| \leqslant \|Az\|_2 \leqslant s_k(A)$$

for all $1 \leqslant k \leqslant n$. □

The following matrix exponential inequality is also interesting.

Theorem 3.20. *For any $A \in \mathbb{C}_{n \times n}$, we have*

$$s(e^A) \prec_{w\text{-}log} e^{s(A)}, \tag{3.23}$$

where $e^{s(A)} = (e^{s_1(A)}, e^{s_2(A)}, \ldots, e^{s_n(A)})$.

Proof. For $1 \leqslant k \leqslant n$, we have

$$\prod_{i=1}^k s_i(e^A) \leqslant \prod_{i=1}^k [\lambda_i(e^{A^*+A})]^{1/2} \qquad \text{(by Theorem 3.16)}$$

$$= \prod_{i=1}^k \lambda_i(e^{(A^*+A)/2})$$

$$= \prod_{i=1}^k e^{\lambda_i(\operatorname{Re} A)}$$

$$\leqslant \prod_{i=1}^k e^{s_i(A)}. \qquad \text{(by Theorem 3.19)}$$

Thus (3.23) is established. □

Notes and References. The proof of Theorem 3.19 is adopted from [Bha97, Proposition III.5.1], while the result was originally from [FH55, p.114–115].

3.4 Extensions to Lie Groups

Let us recall some notations in the context of Lie groups, as in Chapter 2. Let G be a noncompact connected semisimple Lie group with Lie algebra $\mathfrak{g}$. Let $\mathfrak{g} = \mathfrak{k} \oplus \mathfrak{p}$ be a fixed Cartan decomposition of $\mathfrak{g}$, with θ the corresponding Cartan involution. For each $X \in \mathfrak{g}$, write $X = X_\mathfrak{k} + X_\mathfrak{p}$ with $X_\mathfrak{k} \in \mathfrak{k}$ and $X_\mathfrak{p} \in \mathfrak{p}$. Let Θ be the derived Cartan involution of G, let $P = \exp \mathfrak{p}$, and let $G = PK$ and $G = KP$ denote the corresponding Cartan decompositions. For each $g \in G$, denote $g^* = \Theta(g^{-1})$.

Let $\mathfrak{a}$ be any maximal abelian subspace of $\mathfrak{p}$ and pick a closed Weyl chamber $\mathfrak{a}_+$ of $\mathfrak{a}$. Let $A = \exp \mathfrak{a}$ and $A_+ = \exp \mathfrak{a}_+$. Let W be the Weyl group of $(\mathfrak{g}, \mathfrak{a})$. Let $\mathfrak{g} = \mathfrak{k} \oplus \mathfrak{a} \oplus \mathfrak{n}$ and $G = KAN$ be the corresponding Iwasawa decompositions.

Let $\prec$ denote the Kostant preorder as given in Definition 2.20.

Let $\mathfrak{g}_\mathbb{C} = \mathfrak{g} + i\mathfrak{g}$ be the complexification of $\mathfrak{g}$. Since $\mathfrak{g} = \mathfrak{k} \oplus \mathfrak{p}$ is a Cartan decomposition, according to Section 2.5,

$$\mathfrak{g}_\mathbb{C} = \mathfrak{u} \oplus i\mathfrak{u} = (\mathfrak{k} + i\mathfrak{p}) \oplus (\mathfrak{p} + i\mathfrak{k})$$

is a Cartan decomposition of $\mathfrak{g}_\mathbb{C}$, where $\mathfrak{u} = \mathfrak{k} + i\mathfrak{p}$ is a compact real form of $\mathfrak{g}_\mathbb{C}$.

Let $\pi : G \to \operatorname{Aut} V$ be any irreducible representation π of G, and let $d\pi : \mathfrak{g} \to \operatorname{End} V$ be the induced representation of $\mathfrak{g}$ (that is, $d\pi$ is the differential of π at the identity of G). So by (2.1) we have

$$\exp \circ \, d\pi = \pi \circ \exp, \qquad (3.24)$$

where the exponential function on the left is $\exp : \operatorname{End} V \to \operatorname{Aut} V$ and the one on the right is $\exp : \mathfrak{g} \to G$.

Because $\mathfrak{u}$ is a compact real form of $\mathfrak{g}_\mathbb{C}$, there exists a unique (up to scalar) inner product $\langle \cdot, \cdot \rangle$ on V such that $d\pi(X)$ is skew-Hermitian for all $X \in \mathfrak{u}$ and $d\pi(Y)$ is Hermitian for all $Y \in i\mathfrak{u}$, and hence $\pi(k)$ is unitary for $k \in K$ and $\pi(p)$ is positive definite for $p \in P$ by (3.24). We will assume that V is given this inner product.

Now if $g = pk$ with $p \in P$ and $k \in K$, we have

$$\pi(g^*) = (\pi(g))^*, \qquad (3.25)$$

where $(\pi(g))^*$ denotes the adjoint operator of $\pi(g)$, because

$$\pi(g^*) = \pi(k^{-1}p) = (\pi(k))^{-1}\pi(p) = (\pi(k))^*(\pi(p))^* = (\pi(p)\pi(k))^* = (\pi(g))^*.$$

Another result of the existence of an inner product on V is: the operator norm $\|\pi(g)\|$ is well defined for any $g \in G$. So we have $\rho(\pi(g)) \leqslant \|\pi(g)\|$ and

$$\|\pi(fg)\| \leqslant \|\pi(f)\| \cdot \|\pi(g)\|, \qquad \forall f, g \in G.$$

To derive Theorem 3.25, an extension of the Golden-Thompson inequality in the strongest form of Theorem 3.5 (1), we need several preliminary results. The first one is a variation of Theorem 2.21 under some special circumstance ([Kos73, Proposition 4.3]).

Theorem 3.21. (Kostant) *Let $g \in G$ and $p \in P$. If for every irreducible representation π of G,*
$$\|\pi(g)\| \leq \|\pi(p)\|,$$
then we have $g \prec p$. In particular,
$$kpv \prec p, \qquad \forall\, k, v \in K. \tag{3.26}$$

Proof. Let $\pi : G \to \operatorname{End} V$ be any irreducible representation. Since $\pi(p)$ is positive definite for $p \in P$, we have $\rho(\pi(p)) = \|\pi(p)\|$. By assumption,
$$\rho(\pi(g)) \leq \|\pi(g)\| \leq \|\pi(p)\| = \rho(\pi(p)),$$
for every irreducible representation π of G. Therefore, $g \prec p$ by Theorem 2.21.

If $k, v \in K$, then $\pi(k)$ and $\pi(v)$ are unitary, and hence $\|\pi(k)\| = \|\pi(v)\| = 1$. It follows that
$$\|\pi(kpv)\| \leq \|\pi(k)\| \cdot \|\pi(p)\| \cdot \|\pi(v)\| = \|\pi(p)\|,$$
which implies that $kpv \prec p$. $\square$

In the language of matrices, (3.26) amounts to saying that if $A \in \mathbb{P}_n$ and $X \in \mathrm{GL}_n(\mathbb{C})$ has the same singular values with A (counting multiplicity), then
$$|\lambda(X)| \prec_{\log} |\lambda(A)| = s(A) = s(X).$$
In other words, (3.26) is an extension of Theorem 1.17 in Lie groups.

Theorem 3.22. *Let $g \in G$. Then we have*
$$g^2 \prec g^* g. \tag{3.27}$$
More generally,
$$g^{2m} \prec (g^* g)^m, \qquad \forall\, m \in \mathbb{N}. \tag{3.28}$$

Proof. By (3.25), $\pi(g^*) = (\pi(g))^*$ and so $\pi(g)^* \pi(g)$ is positive definite. Then by the properties of operator norm, we have
$$\|\pi(g^2)\| = \|(\pi(g))^2\| \leq \|\pi(g)\|^2 = \|\pi(g)^* \pi(g)\| = \|\pi(g^* g)\|.$$
Since $g^* g \in P$, we see that (3.27) is valid by Theorem 3.21.

For the general case (3.28), one may apply a similar argument:
$$\|\pi(g^{2m})\| = \|(\pi(g))^{2m}\| \leq \|\pi(g)\|^{2m} = \|\pi(g)^* \pi(g)\|^m = \|\pi(g^* g)^m\|.$$
Combining (3.27) and Theorem 2.23 also yields (3.28). $\square$

Theorem 3.23. *If $p, q \in P$, then*

$$(pq)^{2m} \prec (p^2 q^2)^m, \qquad \forall\, m \in \mathbb{N}, \qquad (3.29)$$

$$(pq)^{2^k} \prec p^{2^k} q^{2^k}, \qquad \forall\, k \in \mathbb{N}. \qquad (3.30)$$

Proof. Let $g = pq$. By Theorem 3.22, we have

$$(pq)^{2m} = g^{2m} \prec (g^* g)^m = (qp^2 q)^m = q(p^2 q^2)^m q^{-1}, \qquad \forall\, m \in \mathbb{N}.$$

Now that $q(p^2 q^2)^m q^{-1}$ and $(p^2 q^2)^m$ are conjugate, it follows that (3.29) is valid. Putting $m = 2^{k-1}$ in (3.29), we get

$$(pq)^{2^k} \prec (p^2 q^2)^{2^{k-1}}, \qquad \forall\, k \in \mathbb{N}.$$

Applying this relation iteratively, we have

$$(pq)^{2^k} \prec (p^2 q^2)^{2^{k-1}} \prec (p^4 q^4)^{2^{k-2}} \prec \cdots \prec p^{2^k} q^{2^k}.$$

This completes the proof. $\square$

The following important result is the Lie product formula ([War83, p.110-120]).

Theorem 3.24. (Lie Product Formula) *If $X, Y \in \mathfrak{g}$, then*

$$\lim_{m \to \infty} (e^{X/m} e^{Y/m})^m = e^{X+Y}. \qquad (3.31)$$

Kostant extended the Golden-Thompson inequality (in the ultimate form of Theorem 3.5) to Lie groups ([Kos73, Theorem 6.3]).

Theorem 3.25. (Kostant) *If $X, Y \in \mathfrak{p}$, then*

$$e^{X+Y} \prec e^X e^Y. \qquad (3.32)$$

Proof. Applying (3.30) with $p = e^{X/2^k}$ and $q = e^{Y/2^k}$, we have

$$(e^{X/2^k} e^{Y/2^k})^{2^k} \prec e^X e^Y.$$

Now because the spectral radius is a continuous function on the space of operators, it follows from Theorem 2.21 that Kostant's preorder is continuous. That is,

$$\lim_{k \to \infty} (e^{X/2^k} e^{Y/2^k})^{2^k} \prec e^X e^Y.$$

But the left-hand side is exactly e^{X+Y} by the Lie product formula (3.31). $\square$

Theorem 3.25 is an extension of Theorem 3.5 (1) for semisimple Lie groups: For $G = \mathrm{SL}_n(\mathbb{C})$ and $\mathfrak{g} = \mathfrak{sl}_n(\mathbb{C})$ with Cartan decompositions in Example 2.5 and Example 2.7, the relation $e^{X+Y} \prec e^X e^Y$ with $X, Y \in \mathfrak{p} = i\mathfrak{su}(n)$ means that $\lambda(e^{X+Y}) \prec_{\log} \lambda(e^X e^Y)$ with Hermitian X and Y.

Kostant also showed the following interesting result ([Kos73, Theorem 6.1]).

Theorem 3.26. (Kostant) *Suppose $f, g \in G$ are hyperbolic. If $f \prec g$, then*

$$\chi_\pi(f) \leqslant \chi_\pi(g),$$

for every irreducible representation π of G, where χ_π is the character of π.

It turns out that the converse of Theorem 3.26 is also true, due to Huang-Kim ([HK10, Theorem 3]).

Theorem 3.27. (Huang-Kim) *Suppose $f, g \in G$ are hyperbolic. Then*

$$f \prec g$$

if and only if $\chi_\pi(f) \leqslant \chi_\pi(g)$ for any irreducible representation π of G.

The following result is an extension of the original Golden-Thompson inequality (Theorem 3.1) to Lie groups.

Theorem 3.28. *If $X, Y \in \mathfrak{p}$, then*

$$\chi_\pi(e^{X+Y}) \leqslant \chi_\pi(e^X e^Y) \tag{3.33}$$

for the character χ_π associated with any finite dimensional representation π of G.

Proof. By Theorem 2.19, both e^{X+Y} and $e^X e^Y$ are hyperbolic elements. So (3.33) follows from Theorem 3.25 and Theorem 3.26. □

Recall that the exponential map $\exp : \mathfrak{p} \to P$ is a diffeomorphism onto P. For each $p \in P$, there exists a unique $X \in \mathfrak{p}$ such that $p = \exp X$. Thus it is well defined that

$$p^r = \exp(rX), \quad \forall r \in \mathbb{R}.$$

Recall Theorem 2.19 that the set of all hyperbolic elements in G is

$$L = P^2 = \{pq : p, q \in P\}.$$

Let $\mathfrak{l}$ denote the set of all real semisimple elements in $\mathfrak{g}$. Since the function $\exp : \mathfrak{l} \to L$ is a bijection, it is well defined for each $h = \exp Y \in L$ that

$$h^r = \exp(rY), \quad \forall r \in \mathbb{R}.$$

Obviously, if $p, q \in P$ and $r > 0$, then

$$p^r q^r \in L \quad \text{and} \quad p^{r/2} q^r p^{r/2} = \left(q^{r/2} p^{r/2}\right)^* \left(q^{r/2} p^{r/2}\right) \in P \subset L.$$

The following result is an extension of Theorem 3.8.

Theorem 3.29. *Let $p, q \in P$. The following statements are true and equivalent.*

(1) $p^r q^r \prec (pq)^r$ for all $0 \leqslant r \leqslant 1$.

(2) $(pq)^r \prec p^r q^r$ for all $r \geqslant 1$.

(3) The function $r \mapsto (p^r q^r)^{1/r}$ is monotonically increasing on $(0, \infty)$ in terms of $\prec$. More precisely,

$$(p^r q^r)^{1/r} \prec (p^t q^t)^{1/t}, \qquad \forall 0 < r < t. \tag{3.34}$$

(4) The function $r \mapsto (p^{1/r} q^{1/r})^r$ is monotonically decreasing on $(0, \infty)$ in terms of $\prec$.

Proof. The proof of the equivalence of (1)–(4) is the same as that in Theorem 3.8. We only show that (3) is true.

Suppose $0 < r < t$. Let $\pi : G \to \operatorname{End} V$ be any irreducible finite dimensional representation. Fix an inner product on V such that $\pi(p)$ and $\pi(q)$ are positive definite. By Theorem 2.21, it suffices to show that

$$\rho(\pi[(p^r q^r)^{1/r}]) \leqslant \rho(\pi[(p^t q^t)^{1/t}]), \tag{3.35}$$

where $\rho(\cdot)$ denotes the spectral radius. Thus we have

$$\begin{aligned}
\rho(\pi[(p^r q^r)^{1/r}]) &= \rho([\pi(p^r q^r)]^{1/r}) \\
&= (\rho[\pi(p^r q^r)])^{1/r} \\
&= (\rho([\pi(p)]^r [\pi(q)]^r))^{1/r} \\
&= (\lambda_1([\pi(p)]^r [\pi(q)]^r))^{1/r} \\
&\leqslant (\lambda_1([\pi(p)]^t [\pi(q)]^t))^{1/t} \qquad \text{(by Theorem 3.7(3))} \\
&= (\rho([\pi(p)]^t [\pi(q)]^t))^{1/t} \\
&= \rho(\pi[(p^t q^t)^{1/t}]).
\end{aligned}$$

Thus the desired result (3.35) is valid. This completes the proof. $\square$

Note that $p^r q^r$ and $p^{r/2} q^r p^{r/2}$ are conjugate and that the order $\prec$ is preserved under conjugation. So Theorem 3.29 can also be formulated for $p^{r/2} q^r p^{r/2}$ below, as an extension of Theorem 3.12.

Theorem 3.30. Let $p, q \in P$. The following statements are true and equivalent.

(1) $p^{r/2} q^r p^{r/2} \prec (p^{1/2} q p^{1/2})^r$ for all $0 \leqslant r \leqslant 1$.

(2) $(p^{1/2} q p^{1/2})^r \prec p^{r/2} q^r p^{r/2}$ for all $r \geqslant 1$.

(3) The function $r \mapsto (p^{r/2} q^r p^{r/2})^{1/r}$ is monotonically increasing on $(0, \infty)$ in terms of $\prec$.

(4) The function $r \mapsto (p^{1/2r} q^{1/r} p^{1/2r})^r$ is monotonically decreasing on $(0, \infty)$ in terms of $\prec$.

The following result is an extension of the original Araki-Lieb-Thirring inequality (Theorem 3.11) in Lie groups.

Theorem 3.31. *If $p, q \in P$, then*

$$\chi_\pi((p^{1/2}qp^{1/2})^{rq}) \leqslant \chi_\pi((p^{r/2}q^r p^{r/2})^q), \qquad \forall q \geqslant 0, r \geqslant 1, \qquad (3.36)$$

for the character χ_π associated with any finite dimensional representation π of G.

Proof. By Theorem 2.19, both $(p^{1/2}qp^{1/2})^{rq}$ and $(p^{r/2}q^r p^{r/2})^q$ are in P and hence are hyperbolic elements. So (3.36) follows from Theorem 3.26 and Theorem 3.30. □

Now we extend the following result (as one of seven equivalent statements in Theorem 3.9) for $A, B \in \mathbb{P}_n$ to Lie groups:

$$[s(A^r B^r)]^{1/r} \prec_{\log} [s(A^t B^t)]^{1/t}, \qquad 0 < r < t. \qquad (3.37)$$

To do so, we first recall Theorem 2.14, an extension of singular value decomposition to the Lie group G. Let $\mathfrak{a}$ be a maximal abelian subspace of $\mathfrak{p}$. Fix a closed Weyl chamber $\mathfrak{a}_+$ of $\mathfrak{a}$. Let $A_+ = \exp \mathfrak{a}_+$. Each element in $\mathfrak{p}$ is K-conjugate to a unique element in $\mathfrak{a}_+$, and each element in P is K-conjugate to a unique element in A_+. Consequently, we have

$$G = KP = KA_+ K.$$

In other words, each $g \in G$ can be written as

$$g = uav,$$

where $u, v \in K$ and $a \in A_+$ is uniquely determined. We denote $a_+(g)$ as the unique element in A_+ given by the KA_+K decomposition.

The diffeomorphism $* : G \to G$ is given by $*(g) = \Theta(g^{-1})$, where Θ is the Cartan involution of G. We denote $*(g) = g^*$ for convenience. Note that

$$a_+(g^* g) = [a_+(g)]^2, \qquad \forall g \in G. \qquad (3.38)$$

This is because if $g = uav$ is a KA_+K decomposition, then

$$a_+(g^* g) = a_+((v^{-1}au^{-1})(uav)) = a_+(v^{-1}a^2 v) = a^2 = [a_+(g)]^2.$$

Then (3.37) can be extended to Lie groups as the following result.

Theorem 3.32. *For all $p, q \in P$, the function $r \mapsto [a_+(p^r q^r)]^{1/r}$ is monotonically increasing on $(0, \infty)$ in terms of $\prec$. In other words,*

$$[a_+(p^r q^r)]^{1/r} \prec [a_+(p^t q^t)]^{1/t}, \qquad \forall 0 < r < t. \qquad (3.39)$$

Proof. Applying (3.38) to $g = p^r q^r$, we see that
$$[a_+(p^r q^r)]^2 = a_+((p^r q^r)^*(p^r q^r)) = a_+(q^r p^{2r} q^r),$$
which is K-conjugate to $q^r p^{2r} q^r \in P$. So $[a_+(p^r q^r)]^{1/r}$ is K-conjugate to $(q^r p^{2r} q^r)^{1/2r}$. Moreover, by Theorem 3.30, we have
$$(q^r p^{2r} q^r)^{1/2r} \prec (q^t p^{2t} q^t)^{1/2t}, \qquad \forall\, 0 < r < t.$$
Because the order $\prec$ is preserved by conjugation, the desired relation (3.39) is valid. $\square$

The following result is a refinement of Theorem 3.22 and an extension of Theorem 3.15.

Theorem 3.33. *For any $g \in G$, we have*
$$g^{2m} \prec (g^m)^* g^m \prec (g^* g)^m, \qquad \forall\, m \in \mathbb{N}. \tag{3.40}$$

Proof. Let $m \in \mathbb{N}$. Let π be any irreducible representation of G. By the properties of operator norm, we have
$$\|\pi(g^{2m})\| \leq \|\pi(g^m)\|^2 \leq \|\pi(g)\|^{2m}.$$
Note that
$$\|\pi(g^m)\|^2 = \|[\pi(g^m)]^* \pi(g^m)\|$$
and that
$$\|\pi(g)\|^{2m} = \|[\pi(g)]^* \pi(g)\|^m = \|\pi(g^* g)\|^m = \|[\pi(g^* g)]^m\| = \|\pi((g^* g)^m)\|,$$
where the second to last equality holds because $\pi(g^* g)$ is positive definite. It follows that
$$\|\pi(g^{2m})\| \leq \|\pi((g^m)^* g^m)\| \leq \|\pi((g^* g)^m)\|. \tag{3.41}$$
Since $(g^m)^* g^m$ and $(g^* g)^m$ are in P, (3.41) yields (3.40) by Theorem 3.21. $\square$

By the Cartan decomposition, each $g \in G$ can be uniquely written as
$$g = kp$$
with $k \in K$ and $p \in P$. We denote by $p = p(g)$ the P-component of $g \in G$. Obviously, $p(g) = (g^* g)^{1/2}$.

The following result is an extension of Theorem 3.16.

Theorem 3.34. *For any $X \in \mathfrak{g}$, we have*
$$e^{X^*} e^X \prec e^{X^* + X}. \tag{3.42}$$

Equivalently,
$$p(e^X) \prec e^{X_\mathfrak{p}}$$
for all $X \in \mathfrak{g}$, where $X_\mathfrak{p} = (X^* + X)/2$ denotes the $\mathfrak{p}$-component of X.

Proof. Applying the relation $(g^m)^* g^m \prec (g^*g)^m$ in Theorem 3.33 to $g = e^{X/m}$, we see that
$$(e^X)^* e^X \prec ((e^{X/m})^* e^{X/m})^m, \quad \forall\, m \in \mathbb{N}.$$
Note that $e^{X^*} = (e^X)^*$ by the naturality of the exponential map. So the above relation means that
$$e^{X^*} e^X \prec (e^{X^*/m} e^{X/m})^m, \quad \forall\, m \in \mathbb{N}.$$

Applying the Lie product formula on the right-hand side yields (3.42).

Now (3.42) is equivalent to
$$[p(e^X)]^2 = (e^X)^* e^X \prec e^{X^*+X} = [e^{X_{\mathfrak{p}}}]^2.$$
Thus $p(e^X) \prec e^{X_{\mathfrak{p}}}$ for all $X \in \mathfrak{g}$. □

Now we combine Theorem 3.33, Theorem 3.34, and Theorem 3.25.

Theorem 3.35. *For any $X, Y \in \mathfrak{g}$, we have*
$$e^{X+Y} \prec p(e^{X+Y}) \prec e^{(X+Y)^*/2} e^{(X+Y)/2} \prec e^{X_{\mathfrak{p}}+Y_{\mathfrak{p}}} \prec e^{X_{\mathfrak{p}}} e^{Y_{\mathfrak{p}}}. \qquad (3.43)$$

In particular,
$$e^X \prec p(e^X) \prec e^{X^*/2} e^{X/2} \prec e^{X_{\mathfrak{p}}}, \quad \forall\, X \in \mathfrak{g}. \qquad (3.44)$$

Proof. Let $X, Y \in \mathfrak{g}$. Then Theorem 3.33 yields
$$e^{X+Y} \prec p(e^{X+Y}) \prec e^{(X+Y)^*/2} e^{(X+Y)/2},$$
Theorem 3.34 implies that
$$e^{(X+Y)^*/2} e^{(X+Y)/2} \prec e^{X_{\mathfrak{p}}+Y_{\mathfrak{p}}},$$
and $e^{X_{\mathfrak{p}}+Y_{\mathfrak{p}}} \prec e^{X_{\mathfrak{p}}} e^{Y_{\mathfrak{p}}}$ follows from Theorem 3.25. □

In the case with $\mathfrak{g} = \mathfrak{sl}_n(\mathbb{C})$, the relation $e^X \prec e^{X_{\mathfrak{p}}}$ in (3.44) means that for any $X \in \mathfrak{sl}_n(\mathbb{C})$,
$$e^{\operatorname{Re}\lambda(X)} = |\lambda(e^X)| \prec_{\log} |\lambda(e^{(X+X^*)/2})| = e^{\lambda(\operatorname{Re} X)},$$
where $\operatorname{Re}\lambda(X) = (\operatorname{Re}\lambda_1(X), \ldots, \operatorname{Re}\lambda_n(X))$ denotes the real part of $\lambda(X)$ and the components of $\lambda(\operatorname{Re} X)$ are called *real singular values* or *additive singular values* of X. It follows that
$$\operatorname{Re}\lambda(X) \prec \lambda(\operatorname{Re} X),$$
which is exactly [Bha97, Proposition III.5.3] by Ky Fan.

Finally, we show two results about normal elements. Recall from Section 2.5 that an element $g \in G$ is said to be *normal* if

$$gg^* = g^*g.$$

If $g = kp$ is the Cartan decomposition of $g \in G$, then g is normal if and only if $kp = pk$. The differential of $* : G \to G$ is also denoted by $* : \mathfrak{g} \to \mathfrak{g}$. It turns out that $* = -\theta$ and thus $\mathfrak{p}$ is the eigenspace of $* : \mathfrak{g} \to \mathfrak{g}$ associated with the eigenvalue 1. Consequently, P is the fixed point set of $* : G \to G$. An element $X \in \mathfrak{g}$ is said to be *normal* if

$$[X^*, X] = 0.$$

Obviously, if $X \in \mathfrak{g}$ is normal, then e^X is normal in G.

The following result generalizes Theorem 3.17 and Theorem 3.25.

Theorem 3.36. *If $X, Y \in \mathfrak{g}$ are normal, then*

$$p(e^{X+Y}) \prec p(e^X)p(e^Y). \tag{3.45}$$

In particular, if $X, Y \in \mathfrak{p}$, then (3.45) reduces to (3.32) in Theorem 3.25.

Proof. Since $X \in \mathfrak{g}$ is normal, we have $e^{X^*+X} = e^{X^*}e^X$ and thus

$$e^{X_{\mathfrak{p}}} = e^{\frac{X^*+X}{2}} = \left(e^{X^*+X}\right)^{1/2} = \left(e^{X^*}e^X\right)^{1/2} = p(e^X). \tag{3.46}$$

According to Theorem 3.35, we have

$$p(e^{X+Y}) \prec e^{X_{\mathfrak{p}}+Y_{\mathfrak{p}}} = p(e^X)p(e^Y).$$

This completes the proof. $\square$

The following result generalizes Theorem 3.18 and Theorem 3.30.

Theorem 3.37. *If $X, Y \in \mathfrak{g}$ are normal, then*

$$p(e^{rX/2})p(e^{rY})p(e^{rX/2}) \prec \left(p(e^{X/2})p(e^Y)p(e^{X/2})\right)^r, \quad \forall\, 0 \leqslant r \leqslant 1, \tag{3.47}$$

$$\left(p(e^{X/2})p(e^Y)p(e^{X/2})\right)^r \prec p(e^{rX/2})p(e^{rY})p(e^{rX/2}), \quad \forall\, r \geqslant 1. \tag{3.48}$$

In particular, if $X, Y \in \mathfrak{p}$, then (3.47) and (3.48) reduce to Theorem 3.30 (1) and (2), respectively.

Proof. Since $X \in \mathfrak{g}$ is normal, by (3.46), we have

$$p(e^X) = e^{X_{\mathfrak{p}}}, \quad p(e^{X/2}) = e^{X_{\mathfrak{p}}/2}, \quad p(e^{rX}) = e^{rX_{\mathfrak{p}}}, \quad p(e^{rX/2}) = e^{rX_{\mathfrak{p}}/2}$$

with $X_{\mathfrak{p}} \in \mathfrak{p}$. Thus (3.47) and (3.48) follow from Theorem 3.30 (1) and (2), respectively. $\square$

Remark 3.38. For each $g \in G$, $a_+(g)$ is the unique element in A_+ that is K-conjugate to $p(g) = (g^*g)^{1/2}$. Since Kostant's pre-order $\prec$ on G is invariant under conjugacy, relations in terms of $\prec$ expressed in the form of $a_+(g)$ can also be formulated in the form of $p(g)$, and vice versa.

Notes and References. Theorem 3.33 and Theorem 3.35 are from [Tam10b]. Theorem 3.36 and Theorem 3.37 are from [Liu17].

Chapter 4

Inequalities for Spectral Norm

4.1 Matrix Inequalities for Spectral Norm 91
4.2 Extensions to Lie Groups 96

Throughout this section, let $\|\cdot\|$ denote the spectral norm on $\mathbb{C}_{n\times n}$, defined by
$$\|A\| = \max_{\|x\|_2=1} \|Ax\|_2 = s_1(A).$$

4.1 Matrix Inequalities for Spectral Norm

The following result (see [Kat61] and [Kit93, p.279]) is a generalization of Cordes inequality, which states that for all $A, B \in \mathbb{P}_n$,
$$\|A^r B^r\| \leqslant \|AB\|^r, \qquad \forall 0 \leqslant r \leqslant 1.$$

Theorem 4.1. (Kato) If $A, B, X \in \mathbb{C}_{n\times n}$ with A and B positive definite, then
$$\|A^r X B^r\| \leqslant \|X\|^{1-r} \|AXB\|^r, \qquad \forall 0 \leqslant r \leqslant 1. \tag{4.1}$$

Note that (4.1) can be reformulated as
$$s_1(A^r X B^r) \leqslant [s_1(X)]^{1-r} [s_1(AXB)]^r, \qquad \forall 0 \leqslant r \leqslant 1. \tag{4.2}$$

In this form, Theorem 4.1 can be further generalized.

Theorem 4.2. If $A, B, X \in \mathbb{C}_{n\times n}$ with A and B positive definite, then
$$s(A^r X B^r) \prec_{\log} [s(X)]^{1-r} [s(AXB)]^r, \qquad \forall 0 \leqslant r \leqslant 1, \tag{4.3}$$

where the product of two vectors on the right-hand side is entry-wise.

Proof. By the properties of compound matrices in Theorem 1.15, we have that for each $k = 1, 2, \ldots, n$,

$$\prod_{i=1}^{k} s_i(A^r X B^r) = s_1(C_k(A^r X B^r))$$
$$= s_1([C_k(A)]^r C_k(X)[C_k(B)]^r)$$
$$\leqslant [s_1(C_k(X))]^{1-r}[s_1(C_k(A)C_k(X)C_k(B))]^r \quad \text{by (4.2)}$$
$$= [s_1(C_k(X))]^{1-r}[s_1(C_k(AXB))]^r$$
$$= \left(\prod_{i=1}^{k} s_i(X)\right)^{1-r} \left(\prod_{i=1}^{k} s_i(AXB)\right)^r$$
$$= \prod_{i=1}^{k} [s_i(X)]^{1-r}[s_i(AXB)]^r.$$

When $k = n$, we have

$$\prod_{i=1}^{n} s_i(A^r X B^r) = |\det(A^r X B^r)|$$
$$= [\det(A)]^r \cdot |\det(X)| \cdot [\det(B)]^r$$
$$= (|\det(X)|^{1-r}) \cdot ([\det(A)]^r \cdot |\det(X)|^r \cdot [\det(B)]^r)$$
$$= |\det(X)|^{1-r} \cdot |\det(AXB)|^r$$
$$= \left(\prod_{i=1}^{n} s_i(X)\right)^{1-r} \left(\prod_{i=1}^{n} s_i(AXB)\right)^r$$
$$= \prod_{i=1}^{n} [s_i(X)]^{1-r}[s_i(AXB)]^r.$$

Therefore, (4.3) is valid. □

The following result ([Kit93, p.283]) follows from Theorem 4.1.

Theorem 4.3. (Kato) *If $A, B, X \in \mathbb{C}_{n \times n}$ with A and B positive definite, then*

$$\|A^r X B^{1-r}\| \leqslant \|AX\|^r \|XB\|^{1-r}, \quad \forall 0 \leqslant r \leqslant 1. \tag{4.4}$$

Proof. Since B is positive definite, so is B^{-1}. Applying (4.1) on A, B^{-1}, and XB, we have

$$\|A^r X B^{1-r}\| = \|A^r(XB)B^{-r}\|$$
$$\leqslant \|XB\|^{1-r}\|A(XB)B^{-1}\|^r$$
$$= \|AX\|^r\|XB\|^{1-r}.$$

Therefore, (4.4) is valid. □

Similarly, (4.4) can be reformulated as

$$s_1(A^r X B^{1-r}) \leqslant [s_1(AX)]^r [s_1(XB)]^{1-r}, \qquad \forall\, 0 \leqslant r \leqslant 1. \tag{4.5}$$

Then (4.5) can be generalized as below. Since the proof is similar to that of Theorem 4.2, we omit it.

Theorem 4.4. *If $A, B, X \in \mathbb{C}_{n \times n}$ with A and B positive definite, then*

$$s(A^r X B^{1-r}) \prec_{\log} [s(AX)]^r [s(XB)]^{1-r}, \qquad \forall\, 0 \leqslant r \leqslant 1, \tag{4.6}$$

where the product of two vectors on the right-hand side is entry-wise.

The following result is also interesting ([McI79] and [BD95, p.121]).

Theorem 4.5. (McIntosh) *For all $A, B, X \in \mathbb{C}_{n \times n}$, we have*

$$\|A^* X B\| \leqslant \|AA^* X\|^{1/2} \|X B B^*\|^{1/2}. \tag{4.7}$$

Proof. Let $A = U(A^*A)^{1/2}$ and $B = V(B^*B)^{1/2}$ be the right polar decompositions of A and B, respectively, where $U, V \in \mathbb{C}_{n \times n}$ are unitary. Then $AA^* = U(A^*A)U^{-1}$ and $BB^* = V(B^*B)V^{-1}$. Noting that the spectral norm $\|\cdot\|$ is unitarily invariant and applying Theorem 4.3 for A^*A, B^*B, $U^{-1}XV$, and $r = 1/2$, we have that

$$\begin{aligned}
\|A^* X B\| &= \|(A^*A)^{1/2}(U^{-1}XV)(B^*B)^{1/2}\| \\
&\leqslant \|A^*A(U^{-1}XV)\|^{1/2} \|(U^{-1}XV)(B^*B)\|^{1/2} \\
&= \|U^{-1}AA^*XV\|^{1/2} \|U^{-1}XBB^*V\|^{1/2} \\
&= \|AA^*X\|^{1/2} \|XBB^*\|^{1/2}.
\end{aligned}$$

Thus (4.7) is valid. □

The inequality (4.7) can be reformulated as

$$s_1(A^* X B) \leqslant [s_1(AA^* X)]^{1/2} [s_1(X B B^*)]^{1/2} \tag{4.8}$$

and can be generalized as the following result, whose proof is similar to that of Theorem 4.2.

Theorem 4.6. *For all $A, B, X \in \mathbb{C}_{n \times n}$,*

$$s(A^* X B) \prec_{\log} [s(AA^* X)]^{1/2} [s(X B B^*)]^{1/2}. \tag{4.9}$$

Remark 4.7. (1) Note that (4.1) and (4.4) are equivalent. To see (4.4) $\Rightarrow$ (4.1), we apply (4.4) on A, B^{-1}, and XB to have

$$\begin{aligned}
\|A^r X B^r\| &= \|A^r (XB) B^{-(1-r)}\| \\
&\leqslant \|A(XB)\|^r \|(XB)B^{-1}\|^{1-r} \\
&= \|X\|^{1-r} \|AXB\|^r.
\end{aligned}$$

(2) Note also that (4.4) and (4.7) are equivalent. We show (4.7) ⇒ (4.4). This is true when $r = 1/2$, and it is trivial when $r = 0$ or $r = 1$. Because of the continuity property of the spectral norm, it suffices to show that (4.7) ⇒ (4.4) when $r = k/2^n$ for all $n \in \mathbb{N}$ and $k = 0, 1, 2, \ldots, 2^n$. We do this by induction. The case with $n = 1$ corresponds to $r = 1/2$ and thus is true. Suppose (4.4) is true for all dyadic rationals with denominator 2^{n-1}. Now any reduced rational $r \in [0, 1]$ with denominator 2^n but not 2^{n-1} is of the form $r = k/2^{n-1} + 1/2^n$ for some $0 \leqslant k < 2^{n-1}$. By (4.7) and the induction hypothesis, we have

$$\|A^r X B^{1-r}\|$$
$$= \|A^{1/2}(A^{k/2^{n-1}} X B^{1-(k+1)/2^{n-1}}) B^{1/2}\|$$
$$\leqslant \|(A^{(k+1)/2^{n-1}} X B^{1-(k+1)/2^{n-1}})\|^{1/2} \|A^{k/2^{n-1}} X B^{1-k/2^{n-1}}\|^{1/2}$$
$$\leqslant (\|AX\|^{(k+1)/2^n} \|XB\|^{1/2-(k+1)/2^n})(\|AX\|^{k/2^n} \|XB\|^{1/2-k/2^n})$$
$$= \|AX\|^{(2k+1)/2^n} \|XB\|^{1-(2k+1)/2^n}$$
$$= \|AX\|^r \|XB\|^{1-r}.$$

(3) As in the proof of Theorem 4.2, the relation (4.3) follows from (4.2). Thus they are equivalent. Similarly, (4.5) ⇔ (4.6) and (4.8) ⇔ (4.9). We then conclude that the relations (4.1)–(4.9) are all equivalent, and that they are generalizations of Theorems 3.6, 3.7, 3.8, 3.9, and 3.12.

(4) The equivalence of Theorem 4.2, Theorem 4.4, and Theorem 4.6 can be extended to Lie groups (see Theorem 4.11). This is possible due to the fact that the singular value vector of a nonsingular matrix corresponds to the a_+-component of a Lie group element in KA_+K decomposition.

(5) Theorem 4.1, Theorem 4.3, and Theorem 4.5 are also true for all unitarily invariant norms. While inequalities for unitarily invariant norms have counterparts in Lie groups (see Chapter 5), this is not the case for the product of unitarily invariant norms of matrices. That is why we state the theorems in this chapter only in terms of the spectral norm, but not unitarily invariant norms.

Recall from Theorem 3.6 that the Cordes inequality asserts that the following inequality holds for all $A, B \in \mathbb{P}_n$:

$$\|AB\|^r \leqslant \|A^r B^r\|, \qquad \forall r \geqslant 1.$$

For general $A \in \mathbb{C}_{n \times n}$, the term A^r does not make sense for all positive $r \in \mathbb{R}$. However, the following result is valid for $\mathbb{N}_n$, the set of $n \times n$ normal matrices.

Theorem 4.8. (Nakamoto) *If $A, B \in \mathbb{N}_n$, then for the spectral norm,*

$$\|(AB)^m\| \leqslant \|AB\|^m \leqslant \|A^m B^m\|, \qquad \forall m \in \mathbb{N}. \qquad (4.10)$$

Proof. Obviously, $\|(AB)^m\| \leqslant \|AB\|^m$, since $\|\cdot\|$ is a matrix norm. Let $A = UP_A$ and $B = VP_B$ be right polar decompositions with $P_A = |A| \in \mathbb{P}_n$, $P_B = |B| \in \mathbb{P}_n$, and unitary $U, V \in \mathbb{U}_n$. Note that $UP_A = P_A U$ and $VP_B = P_B V$ by the normality of A and B. Then for each $m \in \mathbb{N}$,

$$\begin{aligned}\|AB\|^m &= \|UP_A P_B V\|^m \\ &= \|P_A P_B\|^m & (\|\cdot\| \text{ is unitarily invariant}) \\ &\leqslant \|P_A{}^m P_B{}^m\| & \text{(by Theorem 3.6)} \\ &= \|U^m P_A{}^m P_B{}^m V^m\| \\ &= \|(UP_A)^m (VP_B)^m\| \\ &= \|A^m B^m\|.\end{aligned}$$

This completes the proof. $\square$

The following result by A. Horn ([Hor54b]) is a generalization of Theorem 4.8.

Theorem 4.9. (A. Horn) *Let $A, B \in \mathbb{N}_n$ and $m \in \mathbb{N}$. Then*

$$s((AB)^m) \prec_{\log} [s(AB)]^m \prec_{\log} s(A^m B^m). \tag{4.11}$$

Proof. By Theorem 3.15, $s(X^m) \prec_{\log} [s(X)]^m$ for all $X \in \mathbb{C}_{n \times n}$ and all $m \in \mathbb{N}$. Thus

$$s((AB)^m) \prec_{\log} [s(AB)]^m.$$

It remains to show the second log-majorization. Because

$$\lambda_1([(AB)^*(AB)]^m) = [\lambda_1((AB)^*(AB))]^m = [s_1(AB)]^{2m} = \|AB\|^{2m}$$

and

$$\lambda_1((A^m B^m)^*(A^m B^m)) = [s_1(A^m B^m)]^2 = \|A^m B^m\|^2,$$

it follows from (4.10) that

$$\lambda_1([(AB)^*(AB)]^m) \leqslant \lambda_1((A^m B^m)^*(A^m B^m)). \tag{4.12}$$

Now by applying (4.12) on $C_k(A)$ and $C_k(B)$, we have that for all $1 \leqslant k \leqslant n$,

$$\begin{aligned}\prod_{j=1}^k \lambda_j([(AB)^*(AB)]^m) &= \lambda_1(C_k([(AB)^*(AB)]^m)) \\ &= \lambda_1([(C_k(A)C_k(B))^*(C_k(A)C_k(B))]^m) \\ &\leqslant \lambda_1(([C_k(A)]^m[C_k(B)]^m)^*([C_k(A)]^m[C_k(B)]^m)) \\ &= \lambda_1(C_k((A^m B^m)^*(A^m B^m))) \\ &= \prod_{j=1}^k \lambda_j((A^m B^m)^*(A^m B^m)).\end{aligned}$$

Furthermore, taking determinant yields

$$\det([(AB)^*(AB)]^m) = \det([A^*A]^m)\det([B^*B]^m) = \det((A^m B^m)^*(A^m B^m)).$$

Therefore,
$$\lambda([(AB)^*(AB)]^m) \prec_{\log} \lambda((A^m B^m)^*(A^m B^m)),$$

which is equivalent to
$$[s(AB)]^m \prec_{\log} s(A^m B^m).$$

This completes the proof. □

Theorem 4.8 and Theorem 4.9 are not valid for general $A, B \in \mathbb{C}_{n \times n}$. Otherwise, it would be true that for all $B \in \mathbb{C}_{n \times n}$

$$\lambda((B^*B)^m) = s((B^*B)^m) \prec_{\log} s((B^m)^*B^m) = \lambda((B^m)^*B^m),$$

which is a contradiction to Theorem 3.15.

Notes and References. According to [FH90], Theorem 4.8 was first shown by Nakamoto.

4.2 Extensions to Lie Groups

Let the notations be as in Chapter 2 and Section 3.4. More precisely, let G be a noncompact connected semisimple Lie group with Lie algebra $\mathfrak{g}$. Let $\mathfrak{g} = \mathfrak{k} \oplus \mathfrak{p}$ be a fixed Cartan decomposition of $\mathfrak{g}$, with θ the corresponding Cartan involution. Let Θ be the derived Cartan involution of G, let $P = \exp \mathfrak{p}$, and let $G = KP$ denote the corresponding Cartan decomposition. For each $g \in G$, denote $g^* = \Theta(g^{-1})$.

Let $\mathfrak{a}$ be any maximal abelian subspace of $\mathfrak{p}$ and pick a closed Weyl chamber $\mathfrak{a}_+$ of $\mathfrak{a}$. Let $A = \exp \mathfrak{a}$ and $A_+ = \exp \mathfrak{a}_+$. Let W be the Weyl group of $(\mathfrak{g}, \mathfrak{a})$. Let $\mathfrak{g} = \mathfrak{k} \oplus \mathfrak{a} \oplus \mathfrak{n}$ and $G = KAN$ be the corresponding Iwasawa decompositions.

Let $\prec$ denote Kostant's preorder as given in Definition 2.20.

Let $\pi : G \to \operatorname{Aut} V$ be any irreducible representation of G, and let $d\pi : \mathfrak{g} \to \operatorname{End} V$ be the induced representation of $\mathfrak{g}$ (that is, $d\pi$ is the differential of π at the identity of G). Let V be endowed with an inner product so that $d\pi(X)$ is skew-Hermitian for all $X \in \mathfrak{k}$ and $d\pi(Y)$ is Hermitian for all $Y \in \mathfrak{p}$, and $\pi(k)$ is unitary for $k \in K$ and $\pi(p)$ is positive definite for $p \in P$.

Each element in $\mathfrak{p}$ is K-conjugate to a unique element in $\mathfrak{a}_+$ and each element in P is K-conjugate to a unique element in A_+. In other words, if $X \in \mathfrak{p}$, then there exist a unique $Z \in \mathfrak{a}_+$ and some $k \in K$ such that

$$X = \operatorname{Ad} k(Z) \quad \text{or} \quad Z = \operatorname{Ad} k^{-1}(X)$$

and

$$\exp X = \exp(\operatorname{Ad} k(Z)) = k \exp(Z) k^{-1}.$$

Recall Theorem 2.14, which states that $G = KA_+K$ and each $g \in G$ can be written as

$$g = uav,$$

where $a \in A_+$ is uniquely determined by g and $u, v \in K$. We denote a as $a_+(g)$ and call it the A_+-component of g. Indeed, $a_+(g)$ is the unique element in A_+ that is K-conjugate to $p(g) = (g^*g)^{1/2}$. So

$$a_+(g) = a_+(g^*) = [a_+(gg^*)]^{1/2} = [a_+(g^*g)]^{1/2} \tag{4.13}$$

and

$$a_+(ugv) = a_+(g), \quad \forall g \in G, \forall u, v \in K. \tag{4.14}$$

The projection map $a_+ : G \to A_+$ defined by $g \mapsto a_+(g)$ is continuous.

We need the following lemma to extend Theorem 4.2, Theorem 4.4, and Theorem 4.6 to Lie groups.

Lemma 4.10. *Let $h_1, h_2 \in A_+$. Then for any irreducible finite dimensional representation $\pi : G \to \operatorname{Aut}(V)$,*

$$\rho(\pi(h_1 h_2)) = \rho(\pi(h_1))\rho(\pi(h_2)),$$

where $\rho(\cdot)$ denotes the spectral radius.

Proof. Since $\mathfrak{a}$ is a (maximal) abelian subspace of $\mathfrak{p}$, it follows that A is an abelian subgroup of G. Thus $\pi(A)$ is an abelian subgroup of positive definite operators, and the elements of $\pi(A)$ are positive diagonal matrices under an appropriate orthonormal basis (once fixed and for all) of V. For each $H \in \mathfrak{a}$,

$$\exp d\pi(H) = \pi(\exp H) \in \pi(A)$$

and hence $d\pi(H)$ is a real diagonal matrix. So $\rho(\pi(e^H))$ is the exponential of the largest diagonal entry of $d\pi(H)$.

Let $H_1, H_2 \in \mathfrak{a}_+$ such that $h_1 = \exp H_1 \in A_+$ and $h_2 = \exp H_2 \in A_+$. Then

$$\pi(h_1 h_2) = \pi(h_1)\pi(h_2) = \exp d\pi(H_1) \exp d\pi(H_2) = \exp d\pi(H_1 + H_2),$$

because $\mathfrak{a}$ is abelian and $d\pi$ is linear as a representation. To arrive at the conclusion

$$\rho(\pi(h_1 h_2)) = \rho(\pi(h_1))\rho(\pi(h_2)),$$

it suffices to show that the sum of the largest diagonal entries of $d\pi(H_1)$ and $d\pi(H_2)$ is the largest diagonal entry of $d\pi(H_1 + H_2)$. To this end, we will use the theory of highest weights [Hum72, p.108] on finite dimensional irreducible representations of the complex semisimple Lie algebra $\mathfrak{g}_\mathbb{C} = \mathfrak{g} + i\mathfrak{g}$.

Let
$$\mathfrak{g} = (\mathfrak{a} \oplus \mathfrak{m}) \oplus \sum_{\alpha \in \Sigma} \mathfrak{g}_\alpha$$

be the restricted root space decomposition of $\mathfrak{g}$ (see Section 2.6), where $\mathfrak{m}$ is the centralizer of $\mathfrak{a}$ in $\mathfrak{k}$ and Σ is the set of restricted roots of $(\mathfrak{g}, \mathfrak{a})$. Let $\mathfrak{h}$ be the maximal abelian subalgebra of $\mathfrak{g}$ containing $\mathfrak{a}$. Then $\mathfrak{a} = \mathfrak{h} \cap \mathfrak{p}$ and we set $\mathfrak{h}_\mathfrak{k} := \mathfrak{h} \cap \mathfrak{k}$. It is known that $\mathfrak{h}_\mathbb{C} := \mathfrak{h} \oplus i\mathfrak{h}$ is a Cartan subalgebra of $\mathfrak{g}_\mathbb{C}$ [Hel78, p.259]. Let Δ be the set of roots of $(\mathfrak{g}_\mathbb{C}, \mathfrak{h}_\mathbb{C})$ and set $\mathfrak{h}_\mathbb{R} := \sum_{\alpha \in \Delta} \mathbb{R} H_\alpha$, where $H_\alpha \in \mathfrak{h}_\mathbb{C}$ is defined by the restriction to $\mathfrak{h}_\mathbb{C}$ of the Killing form, i.e., $B(H_\alpha, H) = \alpha(H)$ for all $H \in \mathfrak{h}_\mathbb{C}$. Then $\mathfrak{h}_\mathbb{C} = \mathfrak{h}_\mathbb{R} \oplus i\mathfrak{h}_\mathbb{R}$ and $\mathfrak{h}_\mathbb{R} = \mathfrak{a} \oplus i\mathfrak{h}_\mathfrak{k}$. Each root $\alpha \in \Delta$ is real-valued on $\mathfrak{h}_\mathbb{R}$ [Hel78, p.170]. Let $\Delta_\mathfrak{p} \subset \Delta$ be the set of roots which do not vanish identically on $\mathfrak{a}$. It is known that Σ is the set of restrictions of $\Delta_\mathfrak{p}$ to $\mathfrak{a}$ [Hel78, p.263]. Furthermore we can choose a positive root system $\Delta^+ \subset \Delta$ so that $\mathfrak{a}_+$ is in the corresponding Weyl chamber (in $\mathfrak{h}_\mathbb{R}$) [Kos73, p.431], that is, $\alpha(H) \geq 0$ for all $H \in \mathfrak{a}_+$, $\alpha \in \Delta^+$. So any root of Δ^+ restricted to either zero or an element in Σ^+ as a linear functional on $\mathfrak{a}$ [Hel78, p.263].

The diagonal entries of the diagonal operator $d\pi(H)$, $H \in \mathfrak{a}_+ \subset \mathfrak{h}_\mathbb{C}$, are the eigenvalues of $d\pi(H)$ so that they are of the form $\mu(H)$, where μ are the weights of the representation $d\pi$ of $\mathfrak{g}_\mathbb{C}$ [Hum72, p.107-108]. Let $\lambda \in \mathfrak{h}'$ be the highest weight of $d\pi$, where $\mathfrak{h}'$ denotes the dual space of $\mathfrak{h}$. From the representation theory we know that $\lambda - \mu$ is a sum of positive roots, i.e.,

$$\lambda - \mu = \sum_{\alpha \in \Delta^+} k_\alpha \alpha, \quad k_\alpha \in \mathbb{N}.$$

Since the restrictions of the positive roots in Δ^+ to $\mathfrak{a}$ are either zero or elements in Σ^+, we conclude $\lambda(H) \geq \mu(H)$ for all $H \in \mathfrak{a}_+$. Since $\mathfrak{a}_+$ is a cone, $H_1 + H_2 \in \mathfrak{a}_+$. Thus $\lambda(H_1 + H_2) = \lambda(H_1) + \lambda(H_2)$ is the largest diagonal entry (eigenvalue) of the diagonal matrix $d\pi(H_1 + H_2)$, and $\lambda(H_1)$ and $\lambda(H_2)$ are the largest diagonal entries (eigenvalues) of $d\pi(H_1)$ and $d\pi(H_2)$, respectively. □

In terms of Kostant's preorder, the following theorem ([Tam08, Theorem 4.3]) is a uniform extension of Theorem 4.2, Theorem 4.4, and Theorem 4.6.

Theorem 4.11. *Let* $0 \leq r \leq 1$. *The following statements are valid and equivalent.*

$$a_+(a^r g b^r) \prec [a_+(g)]^{1-r}[a_+(agb)]^r, \quad \forall\, a, b \in P, g \in G. \quad (4.15)$$

$$a_+(a^r g b^{1-r}) \prec [a_+(ag)]^r[a_+(gb)]^{1-r}, \quad \forall\, a, b \in P, g \in G. \quad (4.16)$$

$$a_+(a^* g b) \prec [a_+(aa^* g)]^{1/2}[a_+(gbb^*)]^{1/2}, \quad \forall\, a, b, g \in G. \quad (4.17)$$

Proof. We will establish (4.16) and show (4.17) $\Rightarrow$ (4.16) $\Rightarrow$ (4.15) $\Rightarrow$ (4.17) to establish the equivalence among the relations.

To show (4.16), suppose $a, b \in P$ and $g \in G$. Write $g = k_1 a_+(g) k_2$ with $a_+(g) \in A_+$ and $k_1, k_2 \in K$. Let π be any finite dimensional irreducible representation of G. Since elements of $\pi(K)$ are unitary and the spectral norm $\|\cdot\|$ is invariant under unitary equivalence, we have

$$\|\pi(g)\| = \|\pi(k_1)\pi(a_+(g))\pi(k_2)\| = \|\pi(a_+(g))\| = \rho(\pi(a_+(g))).$$

Since elements of $\pi(P)$ are positive definite, $\pi(p^r) = [\pi(p)]^r$ for all $p \in P$. Then we have

$$\begin{aligned}
\rho(\pi(a_+(a^r g b^{1-r}))) &= \|\pi(a^r g b^{1-r})\| \\
&= \|[\pi(a)]^r \pi(g) [\pi(b)]^{1-r}\| \\
&\leqslant \|\pi(a)\pi(g)\|^r \|\pi(g)\pi(b)\|^{1-r} \quad \text{(by (4.1))} \\
&= \|\pi(ag)\|^r \|\pi(gb)\|^{1-r} \\
&= \rho(\pi(a_+(ag)))^r \rho(\pi(a_+(gb)))^{1-r} \\
&= \rho(\pi([a_+(ag)]^r)) \rho(\pi([a_+(gb)]^{1-r})) \\
&= \rho(\pi([a_+(ag)]^r [a_+(gb)]^{1-r})). \quad \text{(by Lemma 4.10)}
\end{aligned}$$

By Theorem 2.21, the relation (4.16) is established.

(4.17) $\Rightarrow$ (4.16). Suppose $a, b \in P$ and $g \in G$. For $r = 0$ or $r = 1$, (4.16) is trivial, and the case for $r = 1/2$ follows directly from (4.17) because $a^* = a$ and $b^* = b$. Because of the continuity of the projection map a_+, it suffices to show that (4.17) $\Rightarrow$ (4.16) when $r = k/2^n$ for all $n \in \mathbb{N}$ and $k = 0, 1, 2, \ldots, 2^n$. We show this by induction. The case $n = 1$ corresponds to $r = 1/2$ and thus is true. Suppose (4.16) is true for all dyadic rationals with denominator 2^{n-1}. Now any reduced rational $r \in [0, 1]$ with denominator 2^n but not 2^{n-1} is of the form $r = k/2^{n-1} + 1/2^n$ for some $0 \leqslant k < 2^{n-1}$. By (4.17) and the induction hypothesis, as well as the fact that A is abelian, we have

$$\begin{aligned}
a_+(a^r g b^{1-r}) &= a_+(a^{1/2^n}(a^{k/2^{n-1}} g b^{1-(k+1)/2^{n-1}}) b^{1/2^n}) \\
&\prec [a_+(a^{(k+1)/2^{n-1}} g b^{1-(k+1)/2^{n-1}})]^{1/2} [a_+(a^{k/2^{n-1}} g b^{1-k/2^{n-1}})]^{1/2} \\
&\prec [a_+(ag)]^{(k+1)/2^n} [a_+(gb)]^{1/2-(k+1)/2^n} [a_+(ag)]^{k/2^n} [a_+(gb)]^{1/2-k/2^n} \\
&= [a_+(ag)]^{(2k+1)/2^n} [a_+(gb)]^{1-(2k+1)/2^n} \\
&= [a_+(ag)]^r [a_+(gb)]^{1-r},
\end{aligned}$$

as desired.

(4.16) $\Rightarrow$ (4.15). Suppose $a, b \in P$ and $g \in G$. Then $b^{-1} \in P$. Now by applying (4.16) on a, b^{-1}, and gb, we have

$$\begin{aligned}
a_+(a^r g b^r) &= a_+(a^r (gb) b^{-(1-r)}) \\
&\prec [a_+(a(gb))]^r [a_+((gb)b^{-1})]^{1-r} \\
&= [a_+(g)]^{1-r} [a_+(agb)]^t,
\end{aligned}$$

where the last equation is true because A is abelian.

(4.15) $\Rightarrow$ (4.17). Suppose $a, b, g \in G$. Write $a^* = kp$, $b^* = k'p'$ according to the Cartan decomposition $G = KP$. Then $a = pk^{-1}$, $b = p'(k')^{-1}$, $aa^* = p^2$, and $bb^* = (p')^2$. By applying (4.15) on p^{-2}, $(p')^2$, and p^2g for $r = 1/2$, we have

$$\begin{aligned}a_+(a^*gb) &= a_+(kpgp'(k')^{-1}) \\ &= a_+(pgp') \quad \text{by (4.14)} \\ &= a_+((p^{-2})^{1/2}(p^2g)((p')^2)^{1/2}) \\ &\prec [a_+(p^2g)]^{1/2}[a_+(p^{-2}p^2g(p')^2)]^{1/2} \\ &= [a_+(aa^*g)]^{1/2}[a_+(gbb^*)]^{1/2}.\end{aligned}$$

Thus (4.17) is valid. $\square$

When g is the identity, (4.15) reduces to the following equivalent form of Theorem 3.32:

$$a_+(p^r q^r) \prec [a_+(pq)]^r, \qquad \forall p, q \in P, \forall 0 \leqslant r \leqslant 1. \tag{4.18}$$

When g is the identity, (4.17) yields that

$$a_+(fg) \prec a_+(f)a_+(g), \qquad \forall f, g \in G. \tag{4.19}$$

In the language of matrices, (4.19) amounts to saying that

$$s(AB) \prec_{\log} s(A)s(B)$$

for nonsingular matrices $A, B \in \mathbb{C}_{n \times n}$. In particular, $\|AB\| \leqslant \|A\| \cdot \|B\|$.

Recall that an element $g \in G$ is said to be normal if $gg^* = g^*g$. The following result is an extension of Theorem 4.9 to Lie groups.

Theorem 4.12. Let $f, g \in G$ be normal. Then

$$a_+((fg)^n) \prec a_+^n(fg) \prec a_+(f^n g^n), \qquad \forall n \in \mathbb{N}.$$

Proof. The first relation follows from the fact that

$$a_+(x^n) \prec a_+^n(x), \qquad \forall x \in G, \forall n \in \mathbb{N},$$

which can be derived from

$$(x^n)^* x^n = (x^*)^n x^n \prec (x^*x)^n, \qquad \forall x \in G, \forall n \in \mathbb{N},$$

by Theorem 3.33.

For the second relation, we note that (4.18) is equivalent to

$$[a_+(pq)]^r \prec a_+(p^r q^r), \qquad \forall p, q \in P, \forall r \geqslant 1. \tag{4.20}$$

Let $f = kp$, $g = k'p'$ be the Cartan decompositions of $f, g \in G = KP$. Since f, g are normal, we have that $kp = pk$ and $k'p' = p'k'$. Then for $n \geqslant 1$,

$$a_+^n(fg) = a_+^n(kpk'p') = a_+^n(kpp'k') = a_+^n(pp')$$

and

$$a_+(f^n g^n) = a_+((kp)^n(k'p')^n) = a_+(k^n p^n p'^n k'^n) = a_+(p^n p'^n).$$

Thus by (4.20) we have

$$a_+^n(fg) = a_+^n(pp') \prec a_+(p^n p'^n) = a_+(f^n g^n)$$

for all $n \geqslant 1$. The proof is completed. $\square$

Notes and References. This section is based on [Tam08] and [Liu17, Theorem 3.6].

Chapter 5

Inequalities for Unitarily Invariant Norms

5.1 Matrix Inequalities for Unitarily Invariant Norms 103
5.2 Extensions to Lie Groups 105

Throughout this chapter, let $|||\cdot|||$ denote a unitarily invariant norm on $\mathbb{C}_{n\times n}$. A function $f : \mathbb{R}^n \to \mathbb{R}$ is called a *symmetric gauge function* if f is a vector norm, if $f(Sx) = f(x)$ for all $x \in \mathbb{R}^n$ and for all permutation matrices $S \in \mathbb{C}_{n\times n}$, and if $f(x) = f(|x|)$ for all $x \in \mathbb{R}^n$. An important characterization of unitarily invariant norms given by von Neumann [vN37] is that a function $f : \mathbb{C}_{n\times n} \to \mathbb{R}$ is a unitarily invariant norm if and only if $f(A)$ is a symmetric gauge function on the singular values of A (see [Bha97, p.91]).

5.1 Matrix Inequalities for Unitarily Invariant Norms

Recall Theorem 3.12 that for $A, B \in \mathbb{P}_n$ and $r \geqslant 1$,

$$\lambda((ABA)^r) \prec_{\log} \lambda(A^r B^r A^r). \tag{5.1}$$

By Theorem 1.7 and (5.1), it follows that

$$\lambda((ABA)^r) \prec_w \lambda(A^r B^r A^r). \tag{5.2}$$

Because singular values and eigenvalues coincide for positive definite matrices, by the Fan Dominance Theorem, (5.2) is equivalent to the following result.

Theorem 5.1. *Let $A, B \in \mathbb{P}$ and $r \geqslant 1$. Then for any unitarily invariant norm $|||\cdot|||$,*

$$|||(ABA)^r||| \leqslant |||A^r B^r A^r|||. \tag{5.3}$$

For $A \in \mathbb{C}_{n\times n}$, let $|A| = (A^*A)^{1/2}$. Audenaert ([Aud08, Proposition 3]) obtained the following generalization of Theorem 5.1.

Theorem 5.2. (Audenaert) *Suppose $A, B \in \mathbb{C}_{n\times n}$ with B Hermitian and $r \geqslant 1$. Then*

$$|||\,|ABA^*|^r\,||| \leqslant |||\,|A|^r |B|^r |A|^r\,|||. \tag{5.4}$$

for all unitarily invariant norms $\||\cdot\||$.

The equivalence of the relations in the following result can be established in a manner similar to that of (5.1)–(5.3).

Theorem 5.3. *Suppose $A, B \in \mathbb{C}_{n\times n}$ with B Hermitian and $r \geqslant 1$. Then the following three relations are valid and are equivalent to (5.4):*

$$\lambda_1(|ABA^*|^r) \leqslant \lambda_1(|A|^r|B|^r|A|^r), \tag{5.5}$$
$$\lambda(|ABA^*|^r) \prec_w \lambda(|A|^r|B|^r|A|^r), \tag{5.6}$$
$$\lambda(|ABA^*|^r) \prec_{\log} \lambda(|A|^r|B|^r|A|^r). \tag{5.7}$$

Proof. Because $|ABA^*|$ and $|A|^r|B|^r|A|^r$ are positive semidefinite, their eigenvalues and singular values coincide. So (5.4) $\Leftrightarrow$ (5.6) by Theorem 1.11, and (5.7) $\Rightarrow$ (5.6) $\Rightarrow$ (5.5) by Theorem 1.7. It remains to show (5.5) $\Rightarrow$ (5.7).

We apply a compound matrix argument. First note that $C_k(|A|) = |C_k(A)|$ by right polar decomposition, since the k-th compound matrix is multiplicative. Thus for all $r \geqslant 1$ and $k = 1, \ldots, n-1$,

$$\prod_{i=1}^{k} \lambda_i(|ABA^*|^r) = \lambda_1(C_k(|ABA^*|^r))$$
$$= \lambda_1(|C_k(ABA^*)|^r)$$
$$= \lambda_1(|C_k(A)C_k(B)C_k^*(A)|^r)$$
$$\leqslant \lambda_1(C_k(|A|^r|B|^r|A|^r))$$
$$= \lambda_1(|C_k(A)|^r|C_k(B)|^r|C_k(A)|^r)$$
$$= \lambda_1([C_k(|A|)]^r[C_k(|B|)]^r[C_k(|A|)]^r)$$
$$= \lambda_1(C_k(|A|^r|B|^r|A|^r))$$
$$= \prod_{i=1}^{k} \lambda_i(|A|^r|B|^r|A|^r).$$

Now ABA^* is Hermitian, since B is Hermitian. Therefore,

$$\prod_{i=1}^{n} \lambda_i(|ABA^*|^r) = |\det(ABA^*)|^r$$
$$= |\det(A)|^r \cdot |\det(B)|^r \cdot |\det(A^*)|^r$$
$$= \det(|A|)^r \cdot \det(|B|)^r \cdot \det(|A|)^r$$
$$= \det(|A|^r) \cdot \det(|B|^r) \cdot \det(|A|^r)$$
$$= \det(|A|^r|B|^r|A|^r)$$
$$= \prod_{i=1}^{n} \lambda_i(|A|^r|B|^r|A|^r).$$

So (5.7) follows. □

Theorem 5.2 in the equivalent form of (5.7) can be extended to Lie groups (see Theorem 5.5).

The following result of Simon ([Sim79, p.95]) is also interesting. See [Bha97, p.253, p.285] for historical remarks.

Theorem 5.4. (Simon) *Let $A, B \in \mathbb{C}_{n \times n}$ be such that the product AB is normal. Then*
$$||| AB ||| \leq ||| BA ||| \tag{5.8}$$
for all unitarily invariant norms $||| \cdot |||$ on $\mathbb{C}_{n \times n}$.

If we assume that $X, Y \in \mathbb{C}_{n \times n}$ are nonsingular with X normal, then (5.8) takes the following equivalent form:
$$||| X ||| \leq ||| YXY^{-1} ||| \tag{5.9}$$
for all unitarily invariant norms $||| \cdot |||$, which can be generalized to the form
$$s(X) \prec_{\log} s(YXY^{-1}). \tag{5.10}$$

This is because
$$s(X) = |\lambda(X)| = |\lambda(YXY^{-1})| \prec_{\log} s(YXY^{-1})$$
by the normality of X and Theorem 1.17.

Theorem 5.4 in the stronger form of (5.10) can be extended to Lie groups (see Theorem 5.6).

5.2 Extensions to Lie Groups

Let the notations be as in Chapter 2, Section 3.4, and Section 4.2. More precisely, let G be a noncompact connected semisimple Lie group with Lie algebra $\mathfrak{g}$. Let $\mathfrak{g} = \mathfrak{k} \oplus \mathfrak{p}$ be a fixed Cartan decomposition of $\mathfrak{g}$, with θ the corresponding Cartan involution. Let Θ be the derived Cartan involution of G, let $P = \exp \mathfrak{p}$, and let $G = KP$ denote the corresponding Cartan decomposition. For each $g \in G$, denote $g^* = \Theta(g^{-1})$.

Let $\mathfrak{a}$ be any maximal abelian subspace of $\mathfrak{p}$ and pick a closed Weyl chamber $\mathfrak{a}_+$ of $\mathfrak{a}$. Let $A = \exp \mathfrak{a}$ and $A_+ = \exp \mathfrak{a}_+$. Let W be the Weyl group of $(\mathfrak{g}, \mathfrak{a})$. Let $\mathfrak{g} = \mathfrak{k} \oplus \mathfrak{a} \oplus \mathfrak{n}$ and $G = KAN$ be the corresponding Iwasawa decompositions.

Let $\prec$ denote Kostant's preorder as given in Definition 2.20.

Let $\pi : G \to \mathrm{Aut}\, V$ be any irreducible representation of G, and let $d\pi : \mathfrak{g} \to \mathrm{End}\, V$ be the induced representation of $\mathfrak{g}$ (that is, $d\pi$ is the differential of π at the identity of G). Let V be endowed with an inner product so that $d\pi(X)$ is skew-Hermitian for all $X \in \mathfrak{k}$ and $d\pi(Y)$ is Hermitian for all $Y \in \mathfrak{p}$, and $\pi(k)$ is unitary for $k \in K$ and $\pi(p)$ is positive definite for $p \in P$.

For each $g \in G$, let $p(g) = (g^*g)^{1/2}$ denote the P-component of g and let $k(g)$ denote the K-component of g for the Cartan decomposition $G = PK$.

For each $g \in G$, let $a_+(g)$ denote the A_+-component of g for the decomposition $G = KA_+K$.

The following result is an extension of Theorem 5.2, in the form of (5.7), with respect to Kostant's preorder $\prec$.

Theorem 5.5. *Suppose $g, h \in G$ and $h^* = h$ and $r \geq 1$. Then we have*

$$[p(ghg^*)]^r \prec [p(g)]^r [p(h)]^r [p(g)]^r. \quad (5.11)$$

Proof. By Theorem 2.21, it suffices to show that

$$\rho(\pi([p(ghg^*)]^r)) \leq \rho(\pi([p(g)]^r [p(h)]^r [p(g)]^r)), \quad (5.12)$$

for all irreducible representations $\pi : G \to \operatorname{Aut} V$, where $\rho(\cdot)$ denotes the spectral radius. Fix once and for all an inner product on V such that $\pi(p) \in \operatorname{Aut} V$ is positive definite for all $p \in P$ and $\pi(k) \in \operatorname{Aut} V$ is unitary for all $k \in K$. For $g \in G$, write $g = kp$ with $k \in K$ and $p \in P$. Then $\pi(g) = \pi(k)\pi(p)$ is the right polar decomposition of $\pi(g)$. Also, $g^* = pk^{-1}$ and thus

$$\pi(g^*) = \pi(p)[\pi(k)]^{-1} = (\pi(k)\pi(p))^* = (\pi(g))^*. \quad (5.13)$$

Thus we have

$$|\pi(g)| = ([\pi(g)]^*\pi(g))^{1/2} = \pi(p) = \pi(p(g)). \quad (5.14)$$

From (5.14) and the fact the $\pi(p(g))$ is positive definite, we have

$$\|\pi(g)\| = \||\pi(g)|\| = \|\pi(p(g))\| = \rho(\pi(p(g))), \quad (5.15)$$

where $\|\cdot\|$ denotes the spectral norm on $\operatorname{End} V$. Thus

$$\begin{aligned}
\rho(\pi([p(ghg^*)]^r)) &= \|\pi([p(ghg^*)]^r)\| \\
&= \|[\pi(p(ghg^*))]^r\| \\
&= \||\pi(ghg^*)|^r\| \quad \text{by (5.14)} \\
&= \||\pi(g)\pi(h)[\pi(g)]^*|^r\| \\
&\leq \||\pi(g)|^r|\pi(h)|^r|\pi(g)|^r\| \quad \text{by (5.3)} \\
&= \|[\pi(p(g))]^r[\pi(p(h))]^r[\pi(p(g))]^r\| \quad \text{by (5.14)} \\
&= \|\pi([p(g)]^r)\pi([p(h)]^r)\pi([p(g)]^r)\| \\
&= \|\pi([p(g)]^r[p(h)]^r[p(g)]^r)\| \\
&= \rho(\pi([p(g)]^r[p(h)]^r[p(g)]^r)).
\end{aligned}$$

Thus (5.12) is established. $\square$

Now as an extension of Theorem 5.4, in the form of (5.10), the following theorem asserts that a normal element is the "smallest" in its conjugacy class.

Theorem 5.6. *If $g \in G$ is normal, then for all $h \in G$,*
$$p(g) \prec p(hgh^{-1}).$$

Proof. Let $\pi : G \to \operatorname{Aut} V$ be any finite dimensional irreducible representation. Since $g \in G$ is normal, we have
$$\pi(g)[\pi(g)]^* = \pi(g)\pi(g^*) = \pi(gg^*) = \pi(g^*g) = \pi(g^*)\pi(g) = [\pi(g)]^*\pi(g).$$

So $\pi(g)$ is a normal operator on V, and hence the spectral radius and the spectral norm of $\pi(g)$ are the same. If $\|\cdot\|$ denotes the spectral norm, then
$$\begin{aligned}
\rho(\pi(p(g))) &= \||\pi(g)|\| \quad \text{by (5.14)} \\
&= \|\pi(g)\| \\
&= \rho(\pi(g)) \\
&= \rho(\pi(g)\pi(h)\pi(g^{-1})) \\
&\leqslant \|\pi(g)\pi(h)\pi(g^{-1})\| \\
&= \|\pi(ghg^{-1})\| \\
&= \||\pi(ghg^{-1})|\| \\
&= \|\pi(p(ghg^{-1}))\| \quad \text{by (5.14)} \\
&= \rho(\pi(p(ghg^{-1}))).
\end{aligned}$$

By Theorem 2.21, we have the desired result. $\square$

Notes and References. This section is based on [LT14].

Chapter 6

Inequalities for Geometric Means

6.1	Matrix Inequalities for Geometric Means	109
6.2	Symmetric Spaces	111
6.3	Extensions to Lie Groups	114
6.4	Geodesic Triangles in Symmetric Spaces	114

The Golden-Thompson trace inequality is complemented by the fact that for all $X, Y \in \mathbb{H}_n$ and $t \in [0,1]$,
$$\operatorname{tr} e^X \#_t e^Y \leqslant \operatorname{tr} e^{X+Y} \leqslant \operatorname{tr} e^X e^Y,$$
where $e^X \#_t e^Y$ is the t-geometric mean of e^X and e^Y (defined in Section 6.1). Recently, the t-geometric mean has been gaining intensive interest (e.g., see [And79, AH94, ALM04, Bha07, Bha13, BH06, HP93, LL01, Lim12, Moa05, Yam12] and the references therein), partially because of its connection with Riemannian geometry. More precisely, $\mathbb{P}_n$ may be equipped with a suitable Riemannian metric so that the curve $\gamma(t) = A \#_t B$, $0 \leqslant t \leqslant 1$, is the unique geodesic joining A and B in $\mathbb{P}_n$ ([Bha07, p.205]).

In this chapter, we will first summarize some inequalities for t-geometric means and then extend them to Lie groups in terms of Kostant's preorder. The t-geometric means can also be defined on symmetric spaces of noncompact type. Such spaces have nonpositive curvature, with a very interesting convexity property of geodesic triangle. Although $\mathbb{P}_n$ is not a symmetric space of noncompact type, the space $\mathbb{P}_n^1$ of matrices in $\mathbb{P}_n$ of determinant 1 is.

6.1 Matrix Inequalities for Geometric Means

Let $\mathbb{P}_n$ be the set of $n \times n$ positive definite matrices over $\mathbb{C}$. Suppose $A, B \in \mathbb{P}_n$. The *geometric mean* of A and B is defined as
$$A \# B = A^{1/2} \left(A^{-1/2} B A^{-1/2} \right)^{1/2} A^{1/2}. \tag{6.1}$$
More generally, for $t \in [0,1]$, the *t-geometric mean* of A and B is
$$A \#_t B = A^{1/2} \left(A^{-1/2} B A^{-1/2} \right)^t A^{1/2}. \tag{6.2}$$

Obviously, we have the following three properties of the t-geometric mean:

(1) $A\#_t B = B\#_{1-t} A$;
(2) $(A\#_t B)^{-1} = A^{-1}\#_t B^{-1}$;
(3) $A\#_t B = A^{1-t} B^t$ when $AB = BA$.

The following interesting result is given in [AH94].

Theorem 6.1. (Ando-Hiai) *Let $A, B \in \mathbb{P}_n$ and $t \in [0, 1]$. Then the following relations are equivalent and valid:*

$$\lambda(A^r \#_t B^r) \prec_{\log} \lambda((A\#_t B)^r), \qquad \forall r \geqslant 1; \qquad (6.3)$$

$$\lambda((A\#_t B)^r) \prec_{\log} \lambda(A^r \#_t B^r), \qquad \forall 0 < r \leqslant 1; \qquad (6.4)$$

$$\lambda((A^p \#_t B^p)^{1/p}) \prec_{\log} \lambda((A^q \#_t B^q)^{1/q}), \qquad \forall 0 < q < p. \qquad (6.5)$$

Proof. The equivalence of (6.3)–(6.5) is similar to that in Theorem 3.8. By a compound matrix argument, the validity of (6.3) can be reduced to the validity of

$$\lambda_1(A^r \#_t B^r) \leqslant \lambda_1((A\#_t B)^r), \qquad \forall r \geqslant 1,$$

which was shown in [AH94, p.119–120]. $\square$

The following result was proved in [HP93, p.172].

Theorem 6.2. (Hiai-Petz) *If $A, B \in \mathbb{P}_n$ and $t \in [0, 1]$, then*

$$\lim_{r \to 0}(A^r \#_t B^r)^{1/r} = e^{(1-t)\log A + t \log B}. \qquad (6.6)$$

Combining of Theorem 6.1, Theorem 6.2, and Theorem 3.8 yields the following result.

Theorem 6.3. *If $A, B \in \mathbb{P}_n$ and $t \in [0, 1]$, then*

$$\lambda((A^r \#_t B^r)^{1/r}) \prec_{\log} \lambda(e^{(1-t)\log A + t \log B}) \qquad (6.7)$$

$$\prec_{\log} \lambda\left(A^{(1-t)s} B^{ts}\right)^{1/s} \qquad (6.8)$$

$$= \lambda\left(A^{(1-t)s/2} B^{ts} A^{(1-t)s/2}\right)^{1/s}, \qquad (6.9)$$

for all $r > 0$ and $s > 0$.

Let $A, B \in \mathbb{P}_n$ and $t \in [0, 1]$. From Theorem 1.7 and Theorem 1.11 and the above results, it follows that for any unitarily invariant norm $|||\cdot|||$, the function

$$r \mapsto |||(A^r \#_t B^r)^{1/r}|||$$

increases to $|||e^{(1-t)\log A + t \log B}|||$ as r decreases to 0. In particular, the function

$$r \mapsto \operatorname{tr}(A^{1/r} \#_t B^{1/r})^r$$

is monotonically increasing on $(0,\infty)$ to $\operatorname{tr} e^{(1-t)\log A + t\log B}$ as r increases to ∞, which is complementary to Corollary 3.10 (4).

It is known that $\mathbb{P}_n$ is a Riemannian manifold of nonpositive curvature (i.e., all sectional curvatures are $\leqslant 0$). In general, the distance on a Riemannian manifold of nonpositive curvature has a nice convexity property ([Lan99, IX.Theorem 4.3]): For any two geodesics $\alpha(t)$ and $\beta(t)$ with $t \in \mathbb{R}$, the Riemannian distance $d(\alpha(t),\beta(t))$ between $\alpha(t)$ and $\beta(t)$ is a convex function of t. In particular, for two geodesics $\alpha(t)$ and $\beta(t)$ with $\alpha(0) = A$, $\alpha(1) = B$, $\beta(0) = C$, and $\beta(1) = D$, we have

$$d(\alpha(t),\beta(t)) \leqslant (1-t)d(A,C) + td(B,D), \quad 0 \leqslant t \leqslant 1. \tag{6.10}$$

When $C = D$, for any geodesic $\alpha(t)$ joining A to B and not containing C, we have

$$d(\alpha(t),C) \leqslant (1-t)d(A,C) + td(B,C). \tag{6.11}$$

In particular, we have for A and B in $\mathbb{P}_n$,

$$d(A\#_t B, C) \leqslant (1-t)d(A,C) + td(B,C), \tag{6.12}$$

where $d(A,B) = \|\lambda(\log BA^{-1})\|$ (see [Bha07, p.205-206]) and $\|\cdot\|$ is the Euclidean norm (the distance d is denoted as δ_2 in [Bha07]).

6.2 Symmetric Spaces

The reader is referred to [Hel78] for the standard notations and facts on symmetric spaces.

Let G be a connected semisimple Lie group with Lie algebra $\mathfrak{g}$, let $\mathfrak{g} = \mathfrak{k} \oplus \mathfrak{p}$ be a Cartan decomposition of $\mathfrak{g}$, and let θ and Θ be the corresponding Cartan involutions on $\mathfrak{g}$ and on G, respectively. Let K be a compact subgroup of G such that K is the fixed point set of Θ. Then the homogeneous space G/K is a symmetric space of noncompact type.

Let $G = PK$ be the left Cartan decomposition of G. Let $* : G \to G$ be the diffeomorphism defined by $*(g) = \Theta(g^{-1})$. We denote $*(g)$ by g^*. The identity element of G is denoted by 1. Let us consider the following diagram:

$$\begin{array}{c} G \\ \pi \downarrow \;\;\searrow^{\psi} \\ G/K \underset{\phi}{\overset{\phi^{-1}}{\rightleftarrows}} P. \end{array} \tag{6.13}$$

The map $\pi : G \to G/K$ is the natural projection given by
$$\pi(g) = gK.$$
The map $\psi : G \to P$ is defined by
$$\psi(g) = gg^*$$
and is obviously surjective. The map $\phi : P \to G/K$ is defined by
$$\phi(p) = p^{1/2}K.$$
Because $\exp : \mathfrak{p} \to P$ is bijective, the Cartan decomposition $G = PK$ guarantees that ϕ is a diffeomorphism. Therefore, P may be identified with G/K and regarded as a symmetric space of noncompact type. The inverse of ϕ is given by
$$\phi^{-1}(gK) = gg^*.$$
The diagram in (6.13) is obviously commutative: $\phi \circ \psi = \pi$ and $\phi^{-1} \circ \pi = \psi$.

The natural action of G on the symmetric space G/K is given by
$$f \cdot gK = (fg)K, \qquad \forall f \in G. \tag{6.14}$$
Thus, through ψ and ψ^{-1}, the action of G on P is given by
$$f \cdot p = fpf^*, \qquad \forall f \in G, \tag{6.15}$$
because $f \cdot p = \psi^{-1}(f \cdot \phi(p)) = \psi^{-1}(f \cdot p^{1/2}K) = \psi^{-1}((fp^{1/2})K) = fpf^*$.

Recall Theorem 2.19, which states that $L = P^2$ is the set of all hyperbolic elements in G. Let $\mathfrak{l}$ denote the set of all real semisimple elements in $\mathfrak{g}$. The restriction of the exponential map on $\mathfrak{l}$ is then a bijection onto L. According to Theorem 2.16, $X \in \mathfrak{l}$ if and only if $\mathrm{Ad}\, f(X) \in \mathfrak{a}$ for some $f \in G$. Since $\mathfrak{p} = \mathrm{Ad}\, K(\mathfrak{a})$, we have that
$$\mathfrak{l} = \mathrm{Ad}\, G(\mathfrak{a}) = \mathrm{Ad}\, G(\mathfrak{p}).$$

Suppose $r = gK \in G/K$ for some $g \in G$. Let K_r denote the subgroup of G that fixes r. It is not hard to see that
$$K_r = gKg^{-1} = \{gkg^{-1} : k \in K\}.$$
The Lie algebra of K_r is $\mathfrak{k}_r = \mathrm{Ad}\,(g)\mathfrak{k}$. Thus, if $\mathfrak{p}_r$ is the orthogonal complement of $\mathfrak{k}_r$ in $\mathfrak{g}$, then $\mathfrak{p}_r = \mathrm{Ad}\,(g)\mathfrak{p} \subset \mathfrak{l}$ is the tangent space of G/K at r and
$$\mathfrak{g} = \mathfrak{k}_r \oplus \mathfrak{p}_r$$
is another Cartan decomposition. In particular, $\mathfrak{p}$ is the tangent space of G/K at $1K$.

According to [Hel78, IV.Theorem 3.3], the geodesic in G/K emanating from $r = gK$ with tangent vector $X \in \mathfrak{p}_r$ has the form
$$e^{tX} \cdot r = e^{tX}gK = ge^{tY}K, \tag{6.16}$$
where $Y = \mathrm{Ad}\, g^{-1}(X) \in \mathfrak{p}$.

Theorem 6.4. *Suppose $r = p^{1/2}K$ and $s = q^{1/2}K$ with $p, q \in P$. If*
$$\gamma(t) = e^{tX} \cdot r, \qquad X \in \mathfrak{p}_r,$$
is the parametrization of the geodesic in G/K emanating from r such that $\gamma(1) = s$, then the generating tangent vector is $\log(qp^{-1})/2$.

Proof. Note that
$$q^{1/2}K = s = \gamma(1) = e^X \cdot r = e^X p^{1/2}K = p^{1/2} e^{\operatorname{Ad} p^{-1/2}(X)} K,$$
with $\operatorname{Ad} p^{-1/2}(X) \in \mathfrak{p}$. Thus
$$q = q^{1/2}(q^{1/2})^* = (p^{1/2}e^{\operatorname{Ad} p^{-1/2}(X)})(p^{1/2}e^{\operatorname{Ad} p^{-1/2}(X)})^* = p^{1/2}e^{2\operatorname{Ad} p^{-1/2}(X)}p^{1/2}.$$
It follows that
$$p^{-1/2}e^X p^{1/2} = e^{\operatorname{Ad} p^{-1/2}(X)} = (p^{-1/2}qp^{-1/2})^{1/2}.$$
Therefore,
$$e^X = p^{1/2}(p^{-1/2}qp^{-1/2})^{1/2}p^{-1/2} = (p^{1/2}(p^{-1/2}qp^{-1/2})p^{-1/2})^{1/2} = (qp^{-1})^{1/2}.$$
The desired result then follows. □

Through ψ and ψ^{-1}, we see that each geodesic $\gamma(t)$ in P emanating from p has the form
$$\gamma(t) = \phi^{-1}(p^{1/2}e^{tY}K) = p^{1/2}e^{2tY}p^{1/2} \qquad (6.17)$$
for some $Y \in \mathfrak{p}$. The following result is important.

Theorem 6.5. *Let $p, q \in P$. The unique geodesic $\gamma(t)$ joining p and q in P has the following parametrization*
$$\gamma(t) = p^{1/2}\left(p^{-1/2}qp^{-1/2}\right)^t p^{1/2}, \qquad 0 \leqslant t \leqslant 1. \qquad (6.18)$$

Proof. By (6.17), the unique geodesic in P from p (at $t = 0$) to q (at $t = 1$) is given by
$$\gamma(t) = p^{1/2}e^{tY}p^{1/2}$$
for some $Y \in \mathfrak{p}$. Because $q = \gamma(1) = p^{1/2}e^Y p^{1/2}$, it follows that $e^Y = p^{-1/2}qp^{-1/2}$. Therefore, $\gamma(t) = p^{1/2}(p^{-1/2}qp^{-1/2})^t p^{1/2}$, as desired. □

The parametrization (6.18) has the same form as the t-geometric mean (6.2) on $\mathbb{P}_n$. It is then natural to define the t-geometric mean of $p, q \in P$ as
$$p \#_t q = p^{1/2}\left(p^{-1/2}qp^{-1/2}\right)^t p^{1/2}, \qquad 0 \leqslant t \leqslant 1.$$

Many properties of the t-geometric mean on $\mathbb{P}_n$ can be extended to P. For examples, $p \#_t q = q \#_{1-t} p$ and $(p \#_t q)^{-1} = p^{-1} \#_t q^{-1}$ for all $0 \leqslant t \leqslant 1$.

6.3 Extensions to Lie Groups

Let the notations be as in Chapter 2, Section 3.4, Section 4.2, and Section 5.2.

The following result is an extension of Theorem 6.1 to Lie groups.

Theorem 6.6. *Let $X, Y \in \mathfrak{p}$ and $t \in [0,1]$. Then the following relations are equivalent and valid:*

$$e^{rX} \#_t e^{rY} \prec (e^X \#_t e^Y)^r, \qquad \forall\, r \geq 1, \tag{6.19}$$

$$(e^X \#_t e^Y)^r \prec e^{rX} \#_t e^{rY}, \qquad \forall\, 0 < r \leq 1, \tag{6.20}$$

$$(e^{pX} \#_t e^{pY})^{1/p} \prec (e^{qX} \#_t e^{qY})^{1/q}, \qquad \forall\, 0 < q < p. \tag{6.21}$$

Proof. The fact that (6.19) is valid follows from Theorem 2.21 and Theorem 6.1. The proof of the equivalence of (6.19)–(6.21) is similar to that of Theorem 3.6. □

Similarly, the following result is an extension of Theorem 6.2 to G.

Theorem 6.7. *If $X, Y \in \mathfrak{p}$ and $t \in [0,1]$, then*

$$\lim_{r \to 0} (e^{rX} \#_t e^{rY})^{1/r} = e^{(1-t)X + tY}. \tag{6.22}$$

The following result, which follows from Theorem 6.6 and Theorem 6.7, is an extension of Theorem 6.3 to G.

Theorem 6.8. *If $X, Y \in \mathfrak{p}$ and $t \in [0,1]$, then*

$$(e^{rX} \#_t e^{rY})^{1/r} \prec e^{(1-t)X + tY} \tag{6.23}$$

$$\prec \left(e^{(1-t)sX} e^{tsY}\right)^{1/s} \tag{6.24}$$

for all $r > 0$ and $s > 0$.

6.4 Geodesic Triangles in Symmetric Spaces

Recall (2.21), which states that Kostant's preorder $\prec$ on G can be defined on $\mathfrak{g}$ as well, i.e., for $X, Y \in \mathfrak{g}$,

$$X \prec Y \quad \Longleftrightarrow \quad \exp X \prec \exp Y.$$

This preorder for $X, Y \in \mathfrak{l}$ takes the form

$$X \prec Y \iff \operatorname{conv} W \cdot X \subset \operatorname{conv} W \cdot Y.$$

The following result may be regarded as a Lie algebra version of the Golden-Thompson inequality.

Theorem 6.9. *Let $X, Y \in \mathfrak{p}$. Then there exist unique $Z \in \mathfrak{p}$ and $k \in K$ such that $e^X e^Y = e^Z k$. Moreover,*

$$X + Y \prec Z.$$

Proof. The existence and uniqueness of $Z \in \mathfrak{p}$ and $k \in K$ are guaranteed by the Cartan decomposition $G = PK$. By Theorem 3.25 and Theorem 3.21, we see that

$$e^{X+Y} \prec e^X e^Y = e^Z K \prec e^Z.$$

Therefore, $X + Y \prec Z$ by the definition of $\prec$. $\square$

Recall that $\mathfrak{p}_r \subset \mathfrak{l}$ is the tangent space of G/K at r. Because any two points of G/K can be joined by a unique geodesic, the exponential map σ_r at r defines a bijection of $\mathfrak{p}_r$ onto G/K. Thus there exists a unique vector

$$X(r, s) \in \mathfrak{p}_r$$

such that $s = \sigma_r(X(r, s))$. From the action of G on G/K, it can be deduced that

$$\operatorname{Ad} f(X(r, s)) = X(f \cdot r, f \cdot s), \qquad \forall f \in G. \tag{6.25}$$

If $r = 1K$ and $s = pK$ for some $p \in P$, then $\mathfrak{p}_r = \mathfrak{p}$ and hence

$$X(r, s) = \log p. \tag{6.26}$$

The following result gives an explicit formula for $X(r, s)$.

Theorem 6.10. *Suppose $r = p^{1/2}K$ and $s = q^{1/2}K$ with $p, q \in P$. Then*

$$X(r, s) = \frac{\log(qp^{-1})}{2}.$$

In particular, $X(s, r) = -X(r, s)$.

Proof. Consider the G-actions on G/K and on P. By (6.14) and (6.15),

$$p^{1/2} \cdot 1K = p^{1/2}K = r \quad \text{and} \quad p^{1/2} \cdot 1 = p^{1/2} 1 p^{1/2} = p.$$

Let $d = p^{-1/2} q p^{-1/2} \in P$ so that $p^{1/2} \cdot d = q$ and

$$p^{1/2} \cdot d^{1/2} K = (p^{1/2} d^{1/2})K = [(p^{1/2} d^{1/2})(p^{1/2} d^{1/2})^*]^{1/2} K = q^{1/2} K = s.$$

Then
$$\begin{aligned}
e^{X(r,s)} &= e^{X(p^{1/2} \cdot 1K, p^{1/2} \cdot d^{1/2}K)} \\
&= e^{\operatorname{Ad} p^{1/2}(X(1K, d^{1/2}K))} \quad &\text{(by (6.25))} \\
&= p^{1/2} e^{X(1K, d^{1/2}K)} p^{-1/2} \\
&= p^{1/2} d^{1/2} p^{-1/2} \quad &\text{(by (6.26))} \\
&= (p^{1/2} d p^{-1/2})^{1/2} \\
&= (qp^{-1})^{1/2}.
\end{aligned}$$

The desired formula for $X(r,s)$ then follows. □

By combining Theorem 6.4 and Theorem 6.10, we see that $X(r,s)$ is exactly the generating tangent vector of the geodesic $\gamma(t)$ in G/K emanating from r such that $\gamma(1) = s$. Similar to the t-geometric mean in P, it is natural to define the t-geometric mean of r and s in G/k by the geodesic arc from r to s:
$$r \#_t s = e^{tX(r,s)} \cdot r, \qquad \forall\, 0 \leqslant t \leqslant 1. \tag{6.27}$$

The following explicit formula for σ_r is then obvious.

Theorem 6.11. *For $r \in G/K$, the exponential map $\sigma_r : \mathfrak{p}_r \to G/K$ is given by*
$$\sigma_r(X) = e^X \cdot r.$$

Recall that the Cartan involution θ and the Killing form B on $\mathfrak{g}$ induce an inner product B_θ given by
$$B_\theta(X, Y) = -B(X, \theta Y), \qquad \forall\, X, Y \in \mathfrak{g}.$$

Note that B_θ and B coincide on $\mathfrak{p}$, since $\theta Y = -Y$ for all $Y \in \mathfrak{p}$. Define
$$\|Y\| = \sqrt{B(Y, Y)}, \qquad \forall\, Y \in \mathfrak{p}.$$

Because $\mathfrak{l} = \operatorname{Ad} G(\mathfrak{p})$ and the Killing form is $\operatorname{Ad} G$ invariant, we define
$$\|X\| = \sqrt{B(X, X)}, \qquad \forall\, X \in \mathfrak{l}.$$

Let $d(r,s)$ be the distance from r to s in G/K. Since $X(r,s)$ is the generating tangent vector of the geodesic $\gamma(t)$ in G/K emanating from r such that $\gamma(1) = s$, we have
$$d(r,s) = \|X(r,s)\|.$$

We now consider the geodesic triangle formed by three arbitrary distinct points $o, r, s \in G/K$. Since G/K is a symmetric space of negative curvature, we have
$$\|X(r,o) + X(o,s)\| \leqslant \|X(r,s)\|,$$

which follows from $X(s,r) = -X(r,s)$ and a general property of spaces of nonpositive curvature in [Lan99, IX.Corollary 3.10]. However, the geodesic arc $\gamma(t) = e^{tX(r,s)} \cdot r$ with $t \in [0,1]$ contains more information than merely its length $\|X(r,s)\|$. The following interesting result ([Kos73, Theorem 7.2]) is more general.

Theorem 6.12. (Kostant) *Let $o, r, s \in G/K$. Then*

$$X(r,o) + X(o,s) \prec X(r,s).$$

Proof. Since the order $\prec$ on G is preserved by conjugation, the order $\prec$ on $\mathfrak{g}$ is preserved by $\operatorname{Ad} G$. By (6.25), we may assume, without loss of generality, that $o = 1K$. Let $X = X(r,o) = -X(o,r) \in \mathfrak{p}$, $Y = X(o,s) \in \mathfrak{p}$, and $Z = X(r,s) \in \mathfrak{p}_r$. Then by Theorem 6.11 and the definition of $X(r,s)$,

$$e^{X(o,s)}o = s = e^{X(r,s)}r = e^{X(r,s)}e^{X(o,r)}o.$$

In other words,

$$e^Y = e^Z e^{-X} k = e^{-X} e^{\operatorname{Ad} e^X(Z)} k$$

for some $k \in K$. Thus

$$e^X e^Y = e^{\operatorname{Ad} e^X(Z)} k.$$

Now that $r = e^{X(o,r)} \cdot o = e^{-X} K$, we have $\operatorname{Ad} e^X(Z) \in \mathfrak{p}$. Thus

$$X + Y \prec \operatorname{Ad} e^X(Z)$$

by Theorem 6.9. It follows that $X + Y \prec Z$, since $\prec$ is preserved by $\operatorname{Ad} G$. □

The following result is very interesting.

Theorem 6.13. *Let $o, r, s \in G/K$. Then*

$$X(o\#_t r, o\#_t s) \prec tX(r,s), \qquad \forall\, 0 \leqslant t \leqslant 1.$$

Proof. Because of (6.25), we may assume that $o = 1K$. Write $r = p^{1/2}K$ and $s = q^{1/2}K$ with $p, q \in P$. By Theorem 6.10 and (6.27), we have

$$X(r,s) = \frac{\log(qp^{-1})}{2}, \quad o\#_t r = p^{t/2}K, \quad o\#_t s = q^{t/2}K,$$

and

$$X(o\#_t r, o\#_t s) = \frac{\log(q^t p^{-t})}{2}.$$

It follows that

$$e^{2X(o\#_t r, o\#_t s)} = q^t p^{-t} \quad \text{and} \quad e^{2tX(r,s)} = (qp^{-1})^t.$$

Since $q^t p^{-t} \prec (qp^{-1})^t$ for $0 \leqslant t \leqslant 1$ by Theorem 3.29, the desired result follows. □

By taking norms in Theorem 6.13, we obtain

$$d(o\#_t r, o\#_t s) \leqslant t d(r,s). \tag{6.28}$$

This is precisely the distance convexity (6.10) when $A = C = o$, $B = r$ and $D = s$. In fact, Theorem 6.13 may be regarded as a stronger form of the distance convexity (6.10) on G/K, because the latter can be derived from (6.28) as follows.

Theorem 6.14. *Let $r, s, r', s' \in G/K$. Then*

$$d(r'\#_t r, s'\#_t s) \leqslant (1-t) d(r', s') + t d(r, s), \qquad \forall\, 0 \leqslant t \leqslant 1.$$

Proof. Consider the geodesic triangle with vertices r', r, s. By (6.28), we have

$$d(r'\#_t r, r'\#_t s) \leqslant t d(r, s).$$

Similarly, consideration of the geodesic triangle with vertices s, r', s' yields

$$d(r'\#_t s, s'\#_t s) = d(s\#_{1-t} r', s\#_{1-t} s') \leqslant (1-t) d(r', s').$$

By the distance triangular inequality, we get

$$d(r'\#_t r, s'\#_t s) \leqslant d(r'\#_t r, r'\#_t s) + d(r'\#_t s, s'\#_t s)$$
$$\leqslant t d(r, s) + (1-t) d(r', s').$$

This completes the proof. $\square$

Now we consider geodesic triangles in P. Let $p \in P$ and $r = p^{1/2} K \in G/K$. Let K_p denote the subgroup of G that fixes p. Then

$$K_p = p^{1/2} K p^{-1/2} = K_r$$

and

$$\mathfrak{k}_p = \mathfrak{k}_r = \operatorname{Ad} p^{1/2} \mathfrak{k} \quad \text{and} \quad \mathfrak{p}_p = \mathfrak{p}_r = \operatorname{Ad} p^{1/2} \mathfrak{p},$$

where $\mathfrak{k}_p$ is the Lie algebra of K_p and $\mathfrak{p}_p$ is the orthogonal complement of $\mathfrak{k}_p$ in $\mathfrak{g}$. Indeed, $\mathfrak{p}_p$ is the tangent space to P at p. The geodesic emanating from p is of the form

$$\gamma(t) = e^{tX} \cdot p$$

for some $X \in \mathfrak{p}_p$. Suppose $\gamma(1) = q$. Then

$$q = e^X \cdot p$$
$$= e^X p (e^X)^*$$
$$= (e^X p^{1/2})(e^X p^{1/2})^*$$
$$= (p^{1/2} e^{\operatorname{Ad} p^{-1/2}(X)})(p^{1/2} e^{\operatorname{Ad} p^{-1/2}(X)})^*$$
$$= p^{1/2} e^{\operatorname{Ad} p^{-1/2}(2X)} p^{1/2}.$$

It follows that
$$X = \frac{\log(qp^{-1})}{2}.$$

Let $p, q \in P$. It is natural to define
$$X(p,q) = \frac{\log(qp^{-1})}{2},$$
which is the generating tangent vector of the geodesic $\gamma(t)$ emanating from p such that $\gamma(1) = q$. The distance from p to q in P is then
$$d(p,q) = \|X(p,q)\| = \frac{\|\log(qp^{-1})\|}{2}.$$

Note that both $p \#_t q$ and $e^{(1-t)\log p + t \log q}$ are curves in P joining p and q. The former is the unique geodesic arc, while the latter is the exponential of the line segment $(1-t)\log p + t\log q$ in the Euclidean space $\mathfrak{p}$. The relation $p \#_t q \prec e^{(1-t)\log p + t \log q}$ of (6.23) is equivalent to
$$\log(p \#_t q) \prec (1-t)\log p + t \log q.$$

In other words, we have
$$X(1, p\#_t q) \leqslant (1-t)X(1,p) + tX(1,q). \tag{6.29}$$

If translated on P, the property (6.25) of the action of G on G/K is then
$$\operatorname{Ad} f(X(p,q)) = X(f \cdot p, f \cdot q), \quad \forall f \in G. \tag{6.30}$$

By combining (6.29) and (6.30), we have
$$X(o, p\#_t q) \leqslant (1-t)X(o,p) + tX(o,q), \quad \forall o, p, q \in P, \tag{6.31}$$
which is equivalent to (6.23). Taking norms on both sides yields
$$d(o, p\#_t q) \leqslant (1-t)d(o,p) + td(o,q), \quad \forall o, p, q \in P.$$

This is precisely (6.12), so (6.23) may be regarded as a stronger form of this special case of distance convexity. Needless to say, (6.23) is related to the preorder of G and thus does not exist in general for Riemannian manifolds of nonpositive curvature.

One may compare (6.31) with [Bha07, Exercise 6.1.13]. Finally, we note that (6.31) is translated to G/K as
$$X(o, r\#_t s) \prec (1-t)X(o,r) + tX(o,s), \quad \forall o, r, s \in G/K.$$

Notes and References. Bhatia's book [Bha07] has a systematic treatment of matrix geometric means. This section is based on [LLT14].

Chapter 7

Kostant Convexity Theorems

7.1	Kostant Linear Convexity Theorem	121
7.2	A Partial Order	122
7.3	Thompson-Sing and Related Inequalities	127
7.4	Some Matrix Results Associated with SO(n) and Sp(n)	130
7.5	Kostant Nonlinear Convexity Theorem	133
7.6	Thompson Theorem on Complex Symmetric Matrices	134

7.1 Kostant Linear Convexity Theorem

We first recall the Kostant linear convexity theorem [Kos73, Theorem 8.2]. Let G be a noncompact connected semisimple Lie group with Lie algebra $\mathfrak{g}$. Let $\mathfrak{g} = \mathfrak{k} \oplus \mathfrak{p}$ be a fixed Cartan decomposition of $\mathfrak{g}$. Let K be the connected subgroup of G with Lie algebra $\mathfrak{k}$. Note that $\mathfrak{p}$ is the orthogonal complement of $\mathfrak{k}$ in $\mathfrak{g}$ with respect to the Killing form. The Killing form is negative definite on $\mathfrak{k}$ and positive definite on $\mathfrak{p}$ and let $\mathfrak{a} \subset \mathfrak{p}$ be a maximal abelian subspace in $\mathfrak{p}$. Let $\pi : \mathfrak{p} \to \mathfrak{a}$ be the orthogonal projection of $\mathfrak{p}$ on $\mathfrak{a}$.

Theorem 7.1. (Kostant Linear Convexity Theorem) *For any $y \in \mathfrak{a}$, let $K \cdot y$ denote the orbit of y under the adjoint action of K and let $W \cdot y$ denote the orbit of y under the action of the Weyl group W of $(\mathfrak{g}, \mathfrak{a})$. Then*

$$\pi(K \cdot y) = \operatorname{conv} W \cdot y.$$

Theorem 7.1 describes the projection onto $\mathfrak{a}$ of the orbit $K \cdot y$, which is "roundish" and "hollow". For example, let us consider SU(2), whose Lie algebra is

$$\mathfrak{su}(2) = \{ A \in \mathfrak{sl}_2(\mathbb{C}) : A^* = -A \} \cong i\mathfrak{su}(2),$$

i.e., the space of 2×2 Hermitian matrices of zero trace:

$$i\mathfrak{su}(2) = \left\{ \begin{pmatrix} x & y + iz \\ y - iz & -x \end{pmatrix} : x, y, z \in \mathbb{R} \right\} \cong \mathbb{R}^3.$$

So, if $A \in i\mathfrak{su}(2)$ has eigenvalues $\pm\lambda$ with $\lambda > 0$, then

$$\text{Ad}(\text{SU}(2))A = \left\{ U \begin{pmatrix} \lambda & 0 \\ 0 & -\lambda \end{pmatrix} U^* : U \in \text{SU}(2) \right\}$$
$$= \left\{ \begin{pmatrix} x & y+iz \\ y-iz & -x \end{pmatrix} : x^2 + y^2 + z^2 = \lambda^2 \right\},$$

which is identified as the sphere in $\mathbb{R}^3$ centered at the origin with radius $\lambda > 0$.

Theorem 7.1 yields the Schur-Horn theorem (Theorem 1.26) when $\mathfrak{g} = \mathfrak{sl}_n(\mathbb{C})$ (and thus when $\mathfrak{g} = \mathfrak{gl}_n(\mathbb{C})$, as the action of unitary similarity is trivial on the center of $\mathfrak{gl}_n(\mathbb{C})$). This is because the Cartan decomposition $\mathfrak{sl}_n(\mathbb{C}) = \mathfrak{su}(n) \oplus i\mathfrak{su}(n)$ is the Cartesian decomposition and one can pick

$$\mathfrak{a} = \{\text{diag}(d_1, \ldots, d_n) : d_i \in \mathbb{R} \text{ for all } 1 \leqslant i \leqslant n \text{ and } \sum_{i=1}^n d_i = 0\} \quad (7.1)$$

as the maximal abelian subspace in $\mathfrak{p}$. The Weyl group W of $(\mathfrak{g}, \mathfrak{a})$ is isomorphic to the symmetric group S_n [Hel78]. Finally, we apply Theorem 1.6 to obtain Theorem 1.26.

The defining inequalities of majorization give the hyperplanes, but the convex hull statements in Theorem 1.6 and Theorem 7.1 are more geometric and revealing.

7.2 A Partial Order

Kostant's preorder $\prec$ on G can be defined on $\mathfrak{g}$ as (2.21). The restriction of this preorder on $\mathfrak{a}$ means that for $x, y \in \mathfrak{a}$,

$$y \prec x \quad \Longleftrightarrow \quad y \in \text{conv}\, W \cdot x$$
$$\Longleftrightarrow \quad \text{conv}\, W \cdot y \subset \text{conv}\, W \cdot x.$$

This induces a partial order on the Weyl orbits $\{W \cdot x : x \in \mathfrak{a}\}$ of the elements of $\mathfrak{a}$.

Now we introduce a natural partial order on $\mathfrak{a}$ in the following way. Pick a closed fundamental Weyl chamber $\mathfrak{a}_+$ in $\mathfrak{a}$. The choice of the fundamental Weyl chamber $\mathfrak{a}_+$ determines the set of simple roots $\Delta = \{\alpha_1, \ldots, \alpha_n\}$, which is a subset of the dual space $\mathfrak{a}^*$ of $\mathfrak{a}$ [Hum72], and vice versa. Moreover,

$$\mathfrak{a}_+ = \{x \in \mathfrak{a} : \alpha_i(x) \geqslant 0 \text{ for all } 1 \leqslant i \leqslant n\},$$

which is a cone. As the Weyl group W operates simply transitively on the set of Weyl chambers, $\mathfrak{a}_+$ is a fundamental domain for this action.

The restriction of the Killing form $B(\cdot,\cdot)$ of $\mathfrak{g}$ on $\mathfrak{a}$ is an inner product $\langle \cdot,\cdot \rangle$, which naturally induces an inner product $(\cdot,\cdot)$ on $\mathfrak{a}^*$ via the natural isomorphism $\mathfrak{a} \to \mathfrak{a}^*$. For each $\alpha \in \Delta$, denote by $x_\alpha \in \mathfrak{a}$ that corresponds to $2\alpha/(\alpha,\alpha) \in \mathfrak{a}^*$ under the isomorphism, i.e., $2\alpha(x_\alpha)/(\alpha,\alpha) = 1$ and $\alpha(x) = 0$ if $x \in \mathfrak{a}$ is not a multiple of x_α.

Let $\mathfrak{a}'_+$ be the dual cone of $\mathfrak{a}_+$, namely,
$$y \in \mathfrak{a}'_+ \iff \langle x, y \rangle \geqslant 0, \quad \forall\, x \in \mathfrak{a}_+.$$

Define a natural partial order $\leqslant$ on $\mathfrak{a}$ by
$$y \leqslant x \iff x - y \in \mathfrak{a}'_+, \quad x,y \in \mathfrak{a}.$$

Note that
$$\mathfrak{a}'_+ = \left\{ \sum_{i=1}^n r_i x_{\alpha_i} : r_i \geqslant 0 \text{ for all } 1 \leqslant i \leqslant n \right\} \subset \mathfrak{a}$$

and
$$\mathfrak{a}^*_+ = \{\lambda \in \mathfrak{a}^* : \lambda(x) \geqslant 0 \text{ for all } x \in \mathfrak{a}'_+\} \subset \mathfrak{a}^*,$$

where $\mathfrak{a}^*_+$ corresponds to $\mathfrak{a}_+$ under the Killing form induced isomorphism $\mathfrak{a} \to \mathfrak{a}^*$. Moreover,
$$(\mathfrak{a}'_+)^* = \left\{ \lambda \in \mathfrak{a}^* : \lambda = \sum_{i=1}^n r_i \alpha_i \text{ with } r_i \geqslant 0 \text{ for all } 1 \leqslant i \leqslant n \right\} \subset \mathfrak{a}^*.$$

The two orderings $\prec$ and $\leqslant$ are not identical on $\mathfrak{a}$, but they coincide on $\mathfrak{a}_+$ [Kos73]. See also [Bou68, Proposition 18, Chapter VI].

Theorem 7.2. *If $x,y \in \mathfrak{a}_+$, then $y \in \operatorname{conv} W \cdot x$ if and only if $x - y \in \mathfrak{a}'_+$.*

For $x \in \mathfrak{a}$, let $\widetilde{x} \in \mathfrak{a}_+$ such that $\widetilde{x} = \omega \cdot x$ for some $\omega \in W$. Then for any $x,y \in \mathfrak{a}$, we have
$$\begin{aligned} y \prec x &\iff y \in \operatorname{conv} W \cdot x \\ &\iff \widetilde{y} \in \operatorname{conv} W \cdot \widetilde{x} \\ &\iff \langle z, \widetilde{x} - \widetilde{y}\rangle \geqslant 0 \text{ for all } z \in \mathfrak{a}_+ \\ &\iff \widetilde{y} \leqslant \widetilde{x}. \end{aligned} \tag{7.2}$$

Example 7.3. Let us return to $\mathfrak{g} = \mathfrak{sl}_n(\mathbb{C})$. Let $\mathfrak{a}$ be as in (7.1). The Weyl group W is the symmetric group S_n, so we can pick
$$\mathfrak{a}_+ = \{x = \operatorname{diag}(x_1,\ldots,x_n) \in \mathfrak{a} : x_1 \geqslant \cdots \geqslant x_n \text{ and } \sum_{i=1}^n x_i = 0\}.$$

So $\widetilde{x} \in \mathfrak{a}_+$ is obtained by rearranging the entries of $x \in \mathfrak{a}_+$ in descending order. The inner product $\langle x,y\rangle = \sum_{i=1}^n x_i y_i$ can be rewritten as
$$\begin{aligned} \langle x,y \rangle = {}&(x_1 - x_2)y_1 + (x_2 - x_3)(y_1 + y_2) + \\ &\cdots + (x_{n-1} - x_n)(y_1 + \cdots + y_{n-1}) + x_n(y_1 + \cdots + y_n). \end{aligned}$$

This shows that the cone $\mathfrak{a}'_+ = \{y \in \mathfrak{a} : \langle x, y \rangle \geqslant 0 \text{ for all } x \in \mathfrak{a}_+\}$ is given by

$$\mathfrak{a}'_+ = \left\{ \operatorname{diag}(y_1, \ldots, y_n) \in \mathfrak{a} : \sum_{i=1}^{k} y_i \geqslant 0 \text{ for all } 1 \leqslant k \leqslant n-1 \text{ and } \sum_{i=1}^{n} y_i = 0 \right\}.$$

By Theorem 1.6, the partial order $\leqslant$ on $\mathfrak{a}_+$ reduces to majorization $\prec$.

By Theorem 7.1 and (7.2), we have

$$y \in \pi(K \cdot x) \quad \Longleftrightarrow \quad \langle z, \widetilde{x} - \widetilde{y} \rangle \geqslant 0 \text{ for all } z \in \mathfrak{a}_+. \tag{7.3}$$

So (7.3) is useful for the computation of the corresponding inequalities. Indeed, we can interpret (7.3) as the inequality version of the Kostant linear convexity theorem.

Let $G(n)$ be the semidirect product of S_n and $(\mathbb{Z}/2)^n$, i.e., $G(n)$ acts on $\lambda \in \mathbb{R}^n$:

$$(\lambda_1, \ldots, \lambda_n) \mapsto (\pm \lambda_{\theta(1)}, \ldots, \pm \lambda_{\theta(n)}),$$

with $\theta \in S_n$ and any choice of signs.

Let $SG(n)$ be the semidirect product of S_n and $(\mathbb{Z}/2)^{n-1}$, i.e., $SG(n)$ acts on $\lambda \in \mathbb{R}^n$:

$$(\lambda_1, \ldots, \lambda_n) \mapsto (\pm \lambda_{\theta(1)}, \ldots, \pm \lambda_{\theta(n)}),$$

with $\theta \in S_n$ and an even number of negative signs.

We now give the descriptions of $\operatorname{conv} G(n)x$ and $\operatorname{conv} SG(n)x$ in terms of inequalities.

Theorem 7.4. *Let $x, y \in \mathbb{R}^n$.*

(1) $y \in \operatorname{conv} G(n)x$ if and only if

$$\sum_{i=1}^{k} |y_i| \leqslant \sum_{i=1}^{k} |x_i|, \qquad \forall\, 1 \leqslant k \leqslant n, \tag{7.4}$$

after rearranging x and y in descending order with respect to their absolute values.

(2) $y \in \operatorname{conv} SG(n)x$ if and only if

$$\sum_{i=1}^{k} |y_i| \leqslant \sum_{i=1}^{k} |x_i|, \qquad \forall\, 1 \leqslant k \leqslant n, \tag{7.5}$$

$$\sum_{i=1}^{n-1} |y_i| - |y_n| \leqslant \sum_{i=1}^{n-1} |x_i| - |x_n|, \tag{7.6}$$

and, in addition, if the total number of negative terms of x and y is odd,

$$\sum_{i=1}^{n} |y_i| \leqslant \sum_{i=1}^{n-1} |x_i| - |x_n|, \tag{7.7}$$

after rearranging x and y in descending order with respect to their absolute values.

Proof. (1) The group $G(n)$ is a Weyl group of type B_n or C_n [Hel78]. The closed fundamental Weyl chamber can be picked as [BtD85, p.220]

$$\mathfrak{a}_+ = \{(z_1, \ldots, z_n) \in \mathbb{R}^n : z_1 \geqslant \ldots \geqslant z_n\}$$

with $\mathfrak{a} = \mathbb{R}^n$. Note that $\tilde{x} = (|x_1|, \ldots, |x_n|)$ and $\tilde{y} = (|y_1|, \ldots, |y_n|)$ after rearranging the entries of x and y in descending order with respect to their absolute values, respectively. Using the technique in Example 7.3, we see that $\langle z, v \rangle \geqslant 0$ for all $z \in \mathfrak{a}_+$ if and only if $\sum_{i=1}^{k} v_i \geqslant 0$ for all $k = 1, \ldots, n$. Hence, $y \in \operatorname{conv} G(n)x$ if and only if $\langle z, \tilde{x} - \tilde{y} \rangle \geqslant 0$ for all $z \in \mathfrak{a}_+$ by (7.2). So, (7.4) follows, as desired.

(2) The group $SG(n)$ is a Weyl group of type D_n and we can pick the closed fundamental Weyl chamber

$$\mathfrak{a}_+ = \{(z_1, \ldots, z_n) \in \mathbb{R}^n : z_1 \geqslant \ldots \geqslant z_{n-1} \geqslant |z_n|\}$$

with $\mathfrak{a} = \mathbb{R}^n$ [BtD85, p.219]. For all $z \in \mathfrak{a}_+$ with $z_n \geqslant 0$, we have $\langle z, v \rangle \geqslant 0$ if and only if $\sum_{i=1}^{k} v_i \geqslant 0$ for all $1 \leqslant k \leqslant n$. For all $z \in \mathfrak{a}_+$ with $z_n \leqslant 0$, we rewrite the inner product:

$$\langle z, v \rangle = (z_1 - z_2)v_1 + (z_2 - z_3)(v_1 + v_2) + \ldots$$
$$+ (z_{n-1} + z_n)(v_1 + \cdots + v_{n-1}) - z_n(v_1 + \cdots + v_{n-1} - v_n).$$

So, $\langle z, v \rangle \geqslant 0$ if and only if

$$\sum_{i=1}^{k} v_i \geqslant 0, \qquad \forall 1 \leqslant k \leqslant n-1,$$

$$\sum_{i=1}^{n-1} v_i - v_n \geqslant 0.$$

As a result, $\langle z, v \rangle \geqslant 0$ for all $z \in \mathfrak{a}$ if and only if

$$\sum_{i=1}^{k} v_i \geqslant 0, \qquad \forall 1 \leqslant k \leqslant n-1, \tag{7.8}$$

$$\sum_{i=1}^{n-1} v_k \geqslant |v_n|. \tag{7.9}$$

After arranging the entries of x and y in decreasing order with respect to their absolute values, respectively, $\widetilde{x} = (|x_1|, \ldots, |x_{n-1}|, \pm|x_n|)$, where the sign is $+$ (respectively, $-$) if the number of negative x's is even (respectively, odd). So does $\widetilde{y}$. Hence we have four combinations, namely,

$$\begin{cases} \widetilde{x} = (|x_1|, \ldots, |x_n|) \\ \widetilde{y} = (|y_1|, \ldots, |y_n|) \end{cases} \tag{7.10}$$

$$\begin{cases} \widetilde{x} = (|x_1|, \ldots, |x_{n-1}|, -|x_n|) \\ \widetilde{y} = (|y_1|, \ldots, |y_{n-1}|, -|y_n|) \end{cases} \tag{7.11}$$

$$\begin{cases} \widetilde{x} = (|x_1|, \ldots, |x_n|) \\ \widetilde{y} = (|y_1|, \ldots, |y_{n-1}|, -|y_n|) \end{cases} \tag{7.12}$$

$$\begin{cases} \widetilde{x} = (|x_1|, \ldots, |x_{n-1}|, -|x_n|) \\ \widetilde{y} = (|y_1|, \ldots, |y_n|). \end{cases} \tag{7.13}$$

The case when the total number of negative terms in x and y is even (odd) corresponds to (7.10) and (7.11) ((7.12) and (7.13)). If (7.10) or (7.11) holds, for $v = \widetilde{x} - \widetilde{y}$, (7.8) and (7.9) amount to

$$\sum_{i=1}^{k} |y_i| \leqslant \sum_{i=1}^{k} |x_i|, \qquad \forall\, 1 \leqslant k \leqslant n,$$

$$\sum_{i=1}^{n-1} |y_i| - |y_n| \leqslant \sum_{i=1}^{n-1} |x_i| - |x_n|,$$

$$\sum_{i=1}^{n} |y_i| \leqslant \sum_{i=1}^{n} |x_i| \qquad \text{(already included)}.$$

If (7.12) or (7.13) holds, then they become

$$\sum_{i=1}^{k} |y_i| \leqslant \sum_{i=1}^{k} |x_i|, \qquad \forall\, 1 \leqslant k \leqslant n,$$

$$\sum_{i=1}^{n-1} |y_i| - |y_n| \leqslant \sum_{i=1}^{n-1} |x_i| - |x_n|,$$

$$\sum_{i=1}^{n} |y_i| \leqslant \sum_{i=1}^{n-1} |x_i| - |x_n|.$$

This completes the proof. $\qquad \square$

7.3 Thompson-Sing and Related Inequalities

Theorem 7.1 is a very nice generalization of the Schur-Horn theorem and is related to the forthcoming Thompson-Sing inequalities, which were established by Thompson and Sing independently.

Motivated by Schur-Horn's result on the eigenvalues and diagonal entries of a Hermitian matrix, Mirsky [Mir60] asked for the relation between the singular values and diagonal entries of a complex matrix. His question was completely answered by Thompson and Sing [Tho77, Sin76]. In the following theorem, (7.14) is weak majorization and the subtracted terms in (7.15) are somewhat surprising. The proofs of Thompson and Sing rely on clever induction arguments.

Theorem 7.5. (Thompson-Sing) *Let* $s_1 \geqslant \cdots \geqslant s_n \geqslant 0$ *and let* $d_1, \ldots, d_n \in \mathbb{C}$. *There exists* $A \in \mathbb{C}_{n \times n}$ *with singular values* $s_1, \ldots, s_n$ *and diagonal entries* $d_1, \ldots, d_n$ *if and only if*

$$\sum_{i=1}^{k} |d_i| \leqslant \sum_{i=1}^{k} s_i, \qquad \forall\, 1 \leqslant k \leqslant n, \tag{7.14}$$

$$\sum_{i=1}^{n-1} |d_i| - |d_n| \leqslant \sum_{i=1}^{n-1} s_i - s_n, \tag{7.15}$$

after rearranging $d_1, \ldots, d_n$ *in descending order with respect to their moduli.*

By the Singular Value Decomposition, (7.14) and (7.15) completely describe the set

$$D_{\mathrm{U}(n)}(S) := \{\mathrm{diag}\, U^*SV : U, V \in \mathrm{U}(n)\},$$

where $S = \mathrm{diag}\,(s_1, \ldots, s_n)$. One may view that the group $\mathrm{U}(n) \otimes \mathrm{U}(n)$ acts on $\mathbb{C}_{n \times n}$ such that $U \otimes V : S \mapsto U^*SV$ so the set $\{U^*SV : U, V \in \mathrm{U}(n)\}$ is the orbit of S under the action.

In addition to Theorem 7.5, Thompson [Tho77] also obtained many other results including the next two theorems concerning the real counterparts, i.e., when $G = \mathrm{SO}(n) \otimes \mathrm{SO}(n)$ and $G = \mathrm{O}(n) \otimes \mathrm{O}(n)$.

Theorem 7.6. (Thompson, $\mathrm{SO}(n) \otimes \mathrm{SO}(n)$) *Let* $s_1 \geqslant \cdots \geqslant s_n \geqslant 0$ *and let* $d_1, \ldots, d_n \in \mathbb{R}$. *There exists* $A \in \mathbb{R}_{n \times n}$ *with nonnegative (nonpositive) determinant having singular values* $s_1, \ldots, s_n$ *and diagonal entries* $d_1, \ldots, d_n$ *if and only if*

$$\sum_{i=1}^{k} |d_i| \leqslant \sum_{i=1}^{k} s_i, \qquad \forall\, 1 \leqslant k \leqslant n, \tag{7.16}$$

$$\sum_{i=1}^{n-1} |d_i| - |d_n| \leqslant \sum_{i=1}^{n-1} s_i - s_n, \tag{7.17}$$

and in addition, if the number of negative terms among d is odd (even, if nonpositive determinant),

$$\sum_{i=1}^{n}|d_i| \leq \sum_{i=1}^{n-1} s_i - s_n, \tag{7.18}$$

after rearranging $d_1,\ldots,d_n$ in descending order with respect to their absolute values.

Theorem 7.6 is a generalization of the result of A. Horn [Hor54a] about special orthogonal matrices. The subtracted term in (7.18) makes the inequality look more surprising than those in Theorem 7.5, but we will see that this leads to a pleasant geometric picture.

Let
$$D_{\mathrm{SO}(n)}(S) = \{\operatorname{diag}(USV) : U, V \in \mathrm{SO}(n)\}.$$

Theorem 7.6 completely describes the sets $D_{\mathrm{SO}(n)}(S)$ and $D_{\mathrm{SO}(n)}(S^-)$, where $S^- = \operatorname{diag}(s_1, \ldots, s_{n-1}, -s_n)$. Let

$$D_{\mathrm{O}(n)}(S) := \{\operatorname{diag}(USV) : U, V \in \mathrm{O}(n)\}$$
$$= D_{\mathrm{SO}(n)}(S) \cup D_{\mathrm{SO}(n)}(S^-).$$

From Theorem 7.6, we immediately have the following result of Thompson [Tho77], which completely describes $D_{\mathrm{O}(n)}(S)$.

Theorem 7.7. (Thompson) *Let $s_1 \geq \cdots \geq s_n \geq 0$ and let $d_1, \ldots, d_n \in \mathbb{R}$. There exists $A \in \mathbb{R}_{n \times n}$ with singular values $s_1, \ldots, s_n$ and diagonal entries $d_1, \ldots, d_n$ if and only if*

$$\sum_{i=1}^{k}|d_i| \leq \sum_{i=1}^{k} s_i, \quad \forall\, 1 \leq k \leq n, \tag{7.19}$$

$$\sum_{i=1}^{n-1}|d_i| - |d_n| \leq \sum_{i=1}^{n-1} s_i - s_n, \tag{7.20}$$

after rearranging $d_1,\ldots,d_n$ in descending order with respect to their absolute values.

Let $G = \mathrm{SO}(n,n)$, the group of matrices in $\mathrm{SL}_{2n}(\mathbb{R})$ which leaves invariant the quadratic form

$$-x_1^2 - \cdots - x_n^2 + x_{n+1}^2 + \cdots + x_{2n}^2.$$

In other words,
$$\mathrm{SO}(n,n) = \{A \in \mathrm{SL}_{2n}(\mathbb{R}) : A^\top I_{n,n} A = I_{n,n}\},$$

where $I_{n,n} = (-I_n) \oplus I_n$. The group $\mathrm{SO}(n,n)$ has two components and hence is not connected. It is also noncompact. It is well known that [Hel78]

$$\mathfrak{g} = \mathfrak{so}_{n,n} = \left\{ \begin{pmatrix} X_1 & Y \\ Y^\top & X_2 \end{pmatrix} : X_1^\top = -X_1,\ X_2^\top = X_2,\ Y \in \mathbb{R}_{n \times n} \right\},$$
$$K = \mathrm{SO}(n) \times \mathrm{SO}(n),$$
$$\mathfrak{k} = \mathfrak{so}(n) \oplus \mathfrak{so}(n),$$
$$\mathfrak{p} = \left\{ \begin{pmatrix} 0 & Y \\ Y^\top & 0 \end{pmatrix} : Y \in \mathbb{R}_{n \times n} \right\},$$
$$\mathfrak{a} = \bigoplus_{1 \leq j \leq n} \mathbb{R}(E_{j,n+j} + E_{n+j,j}),$$

where $E_{i,j}$ is the $2n \times 2n$ matrix whose only nonzero entry is 1 at the (i,j) position. We identify $\mathfrak{a}$ and $\mathbb{R}^n$ in the obvious way.

The projection π sends

$$\begin{pmatrix} U & 0 \\ 0 & V \end{pmatrix}^\top \begin{pmatrix} 0 & S \\ S & 0 \end{pmatrix} \begin{pmatrix} U & 0 \\ 0 & V \end{pmatrix} = \begin{pmatrix} 0 & U^\top SV \\ V^\top SU & 0 \end{pmatrix}$$
$$\mapsto \begin{pmatrix} 0 & \mathrm{diag}\,(U^\top SV) \\ \mathrm{diag}\,(V^\top SU) & 0 \end{pmatrix},$$

where $U, V \in \mathrm{SO}(n)$. The system of real roots of $\mathfrak{so}_{n,n}$ is of type D_n. The Weyl group W can be viewed as the semidirect product $S_n \times (\mathbb{Z}/2)^{n-1}$, and its action on $\mathfrak{a}$ is:

$$\begin{pmatrix} 0 & D \\ D & 0 \end{pmatrix} \in \mathfrak{a}, \qquad (d_1, \ldots, d_n) \mapsto (\pm d_{\sigma(1)}, \ldots, \pm d_{\sigma(n)}),$$

where $D = \mathrm{diag}\,(d_1, \ldots, d_n)$, $\sigma \in S_n$, and the number of negative signs in $(\pm d_{\sigma(1)}, \ldots, \pm d_{\sigma(n)})$ is even. We just proved Theorem 7.6 and Theorem 7.7.

Now we are going to prove the necessary part of the inequalities in Theorem 7.5. Let
$$O_{\mathrm{U}(n)}(S) := \{U^* SV : U, V \in \mathrm{U}(n)\}.$$

Note that

$$O_{\mathrm{U}(n)}(S) = \{U^* SV : U, V \in \mathrm{U}(n),\ \det U^* = \det V\}$$
$$= \{U^* SV : U, V \in \mathrm{U}(n),\ \det U \det V = 1\}.$$

This is because if $A = U^* SV$ with $U, V \in \mathrm{U}(n)$, then one can find $\theta \in \mathbb{R}$ such that $\det(e^{i\theta} U^*)$ and $\det(e^{-i\theta} V)$ are identical.

Let $G = \mathrm{SU}(n,n)$, the subgroup of $\mathrm{SL}_{2n}(\mathbb{C})$ leaving invariant the Hermitian form
$$-|z_1|^2 - \cdots - |z_n|^2 + |z_{n+1}|^2 + \cdots + |z_{2n}|^2.$$

In other words,
$$\mathrm{SU}(n,n) = \{A \in \mathrm{SL}_{2n}(\mathbb{C}) : A^* I_{n,n} A = I_{n,n}\}.$$

The group $\mathrm{SU}(n,n)$ is connected but noncompact. The Lie algebra $\mathfrak{g} = \mathfrak{su}_{n,n}$ is a real form of $\mathfrak{sl}_{2n}(\mathbb{C})$. It is known that [Hel78]

$$\mathfrak{g} = \left\{ \begin{pmatrix} X_1 & Y \\ Y^\top & X_2 \end{pmatrix} : X_1^* = -X_1,\ X_2^* = X_2,\ \mathrm{tr}\, X_1 = \mathrm{tr}\, X_2 = 0,\ Y \in \mathbb{C}_{n \times n} \right\},$$

$$K = S(\mathrm{U}(n) \times \mathrm{U}(n)) = \left\{ \begin{pmatrix} U & 0 \\ 0 & V \end{pmatrix} : U, V \in \mathrm{U}(n),\ \det U \det V = 1 \right\},$$

$$\mathfrak{k} = \mathfrak{s}(\mathfrak{u}(n) \oplus \mathfrak{u}(n)),$$

$$\mathfrak{p} = \left\{ \begin{pmatrix} 0 & Y \\ Y^* & 0 \end{pmatrix} : Y \in \mathbb{C}_{n \times n} \right\},$$

$$\mathfrak{a} = \bigoplus_{1 \leqslant j \leqslant n} \mathbb{R}(E_{j,n+j} + E_{n+j,j}).$$

The projection π sends

$$\begin{pmatrix} U & 0 \\ 0 & V \end{pmatrix}^* \begin{pmatrix} 0 & S \\ S & 0 \end{pmatrix} \begin{pmatrix} U & 0 \\ 0 & V \end{pmatrix} = \begin{pmatrix} 0 & U^*SV \\ V^*SU & 0 \end{pmatrix}$$

$$\mapsto \begin{pmatrix} 0 & \mathrm{Re\,diag}\,(U^*SV) \\ \mathrm{Re\,diag}\,(V^*SU) & 0 \end{pmatrix},$$

where $U, V \in \mathrm{U}(n)$ and $\det U \det V = 1$. The system of the real roots is of type C_n. The Weyl group W can be viewed as the semidirect product $S_n \ltimes (\mathbb{Z}/2)^n$, and its action on $\mathfrak{a}$ is given by

$$\begin{pmatrix} 0 & D \\ D & 0 \end{pmatrix} \in \mathfrak{a}, \qquad (d_1, \ldots, d_n) \mapsto (\pm d_{\sigma(1)}, \ldots, \pm d_{\sigma(n)}),$$

where $D = \mathrm{diag}\,(d_1, \ldots, d_n)$, $\sigma \in S_n$, with no restriction on the signs.

For $s \in \mathbb{R}_+^n$, we denote by $\mathrm{conv}\, G(n)s$ the convex hull of

$$\{(\pm s_{\sigma(1)}, \ldots, \pm s_{\sigma(n)}) : \sigma \in S_n\}.$$

7.4 Some Matrix Results Associated with $\mathrm{SO}(n)$ and $\mathrm{Sp}(n)$

Both $\mathrm{SO}(n)$ and $\mathrm{Sp}(n)$ are compact connected simple Lie groups, so one can consider $\mathfrak{so}_n(\mathbb{C}) = \mathfrak{so}(n) + i\mathfrak{so}(n)$ and $\mathfrak{sp}_n(\mathbb{C}) = \mathfrak{sp}(n) + i\mathfrak{sp}(n)$ and apply Kostant's linear convexity theorem to obtain the following results, which can be found in [Tam97].

Theorem 7.8. (Tam) *Let $A \in \mathbb{R}_{k \times k}$ be skew symmetric. Then the set*

$$D(A) = \{\operatorname{diag} U^\top AU[1, 3, \ldots, 2n-1 | 2, 4, \ldots, 2n] : U \in \operatorname{SO}(k)\}$$

is convex, where $k = 2n$ or $k = 2n+1$. More precisely, if

$$O^\top AO = \begin{cases} \begin{pmatrix} 0 & \lambda_1 \\ -\lambda_1 & 0 \end{pmatrix} \oplus \cdots \oplus \begin{pmatrix} 0 & \lambda_n \\ -\lambda_n & 0 \end{pmatrix} & \text{if } k = 2n \\ \begin{pmatrix} 0 & \lambda_1 \\ -\lambda_1 & 0 \end{pmatrix} \oplus \cdots \oplus \begin{pmatrix} 0 & \lambda_n \\ -\lambda_n & 0 \end{pmatrix} \oplus 0 & \text{if } k = 2n+1 \end{cases}$$

is the canonical form of A under the adjoint action of $\operatorname{SO}(k)$ for some $O \in \operatorname{SO}(k)$, then

(1) $D(A) = \operatorname{conv} G(n)\lambda$ *when* $k = 2n+1$,

(2) $D(A) = \operatorname{conv} SG(n)\lambda$ *when* $k = 2n$,

where $\lambda = (\lambda_1, \ldots, \lambda_n)$.

Proof. The Lie algebra of $\operatorname{SO}(k)$ is the algebra of $k \times k$ real skew symmetric matrices. The maximal torus of $\operatorname{SO}(2n)$ is $T = \operatorname{SO}(2) \times \cdots \times \operatorname{SO}(2)$, while the maximal torus of $\operatorname{SO}(2n+1)$ is $\operatorname{SO}(2) \times \ldots \operatorname{SO}(2) \times 1$. Considering the injection $A \mapsto A \oplus 1$, we can treat $T \subset \operatorname{SO}(2n+1)$ as a maximal torus of $\operatorname{SO}(2n+1)$. The Lie algebra $\mathfrak{t}$ of T is the set

$$\mathfrak{t} = \left\{ \begin{pmatrix} 0 & \beta_1 \\ -\beta_1 & 0 \end{pmatrix} \oplus \cdots \oplus \begin{pmatrix} 0 & \beta_n \\ -\beta_n & 0 \end{pmatrix} : (\beta_1, \ldots, \beta_n) \in \mathbb{R}^n \right\}.$$

Thus we can identify $\mathfrak{t}$ with $\mathbb{R}^n$ by sending

$$\begin{pmatrix} 0 & \beta_1 \\ -\beta_1 & 0 \end{pmatrix} \oplus \cdots \oplus \begin{pmatrix} 0 & \beta_n \\ -\beta_n & 0 \end{pmatrix} \mapsto (\beta_1, \ldots, \beta_n).$$

When $k = 2n+1$, the Weyl group W operates on the torus T and its action is given by

$$(\beta_1, \ldots, \beta_n) \mapsto (\pm \beta_{\theta(1)}, \ldots, \pm \beta_{\theta(n)}), \tag{7.21}$$

where $\theta \in S_n$ and for any choice of signs.

When $k = 2n$, the Weyl group W operates on the torus T and its action is given by (7.21), where $\theta \in S_n$ and the number of negative signs is even. The desired results then follow from the Kostant linear convexity theorem. □

We have the following corollary by Theorem 7.4.

Corollary 7.9. *Let $A \in \mathbb{R}_{k \times k}$ be skew symmetric and let $D(A)$ and $\lambda = (\lambda_1, \ldots, \lambda_k)$ be as in Theorem 7.8.*

(1) When $k = 2n+1$, in terms of inequalities, $(d_1, \ldots, d_n) \in D(A) \subset \mathbb{R}^n$ if and only if (after rearranging d and λ in decreasing order with respect to their absolute values)

$$\sum_{i=1}^{k} |d_i| \leqslant \sum_{i=1}^{k} |\lambda_i|, \quad 1 \leqslant k \leqslant n.$$

(2) When $k = 2n$, in terms of inequalities, $(d_1, \ldots, d_n) \in D(A) \subset \mathbb{R}^n$ if and only if (after rearranging d and λ in decreasing order with respect to their absolute values)

$$\sum_{i=1}^{k} |d_i| \leqslant \sum_{i=1}^{k} |\lambda_i|, \quad 1 \leqslant k \leqslant n,$$

$$\sum_{i=1}^{n-1} |d_i| - |d_n| \leqslant \sum_{i=1}^{n-1} |\lambda_i| - |\lambda_n|,$$

and in addition, if the total number of negative terms among d and λ is odd,

$$\sum_{i=1}^{n} |d_i| \leqslant \sum_{i=1}^{n-1} |\lambda_i| - |\lambda_n|.$$

Corollary 7.10. (Tam) Let $\Lambda = \operatorname{diag} \lambda = \operatorname{diag}(\lambda_1, \ldots, \lambda_n)$.

(1) The set

$$\left\{ \operatorname{diag}(U^\top \Lambda X - V^\top \Lambda W) : \begin{pmatrix} U & W \\ V & X \end{pmatrix} = Z[1, \ldots, 2n | 1, \ldots, 2n], \right.$$
$$\left. Z \in O(2n+1) \right\}$$

is the convex hull of the elements $(\pm \lambda_{\theta(1)}, \ldots, \pm \lambda_{\theta(n)})$, with $\theta \in S_n$ and any choice of signs.

(2) The two sets

$$\left\{ \operatorname{diag}(U^\top \Lambda X - V^\top \Lambda W) : \begin{pmatrix} U & W \\ V & X \end{pmatrix} \in O(2n) \right\}$$

and

$$\{ \operatorname{diag} U^\top \Lambda X : U, X \in O(n) \}$$

are identical and are equal to the union of the two convex sets $\operatorname{conv} SG(n)\theta$ and $\operatorname{conv} SG(n)\hat{\theta}$, where $\hat{\theta} := (\lambda_1, \ldots, \lambda_{n-1}, -\lambda_n)$.

One can establish the inequalities in Corollary 7.10 by applying Theorem 7.4.

Theorem 7.11. (Tam) *Let $C \in \mathfrak{sp}(n)$ with*

$$C = \begin{pmatrix} A & -\overline{B} \\ B & \overline{A} \end{pmatrix} \in \mathbb{C}_{2n \times 2n}, \quad A^* = -A, \quad B^\top = B.$$

The set

$$R(C) = \{\operatorname{diag}(U^*CU[1,\ldots,n|1,\ldots,n]) : U \in \operatorname{Sp}(n)\}$$

is convex. More precisely, if $\pm i\lambda_1, \ldots, \pm i\lambda_n$ (λ_i's are real) are the eigenvalues of C, then $R(C) = \operatorname{conv} G(n)\lambda$.

Proof. Recall that we have a canonical inclusion $\mathrm{U}(n) \to \operatorname{Sp}(n)$, where

$$A \mapsto \begin{pmatrix} A & 0 \\ 0 & \overline{A} \end{pmatrix} = A + 0j.$$

Let $T^n \subset \operatorname{Sp}(n)$ be the image of the torus $\triangle(n) \subset \mathrm{U}(n)$, which is the subgroup of diagonal matrices. Under this inclusion, T^n is a maximal torus in $\operatorname{Sp}(n)$. The Weyl group of $\operatorname{Sp}(n)$ is also $G(n)$. The group $G(n)$ operates on $T^n = \triangle(n) = T(n)$, where we identify these tori using the inclusions [BtD85, p.173]

$$\operatorname{Sp}(n) \supset \mathrm{U}(n) \subset \operatorname{SO}(2n).$$

The desired result follows from Kostant's linear convexity theorem. □

One can describe the necessary and sufficient conditions in Theorem 7.11 in terms of inequalities via Theorem 7.4(1).

7.5 Kostant Nonlinear Convexity Theorem

The Kostant nonlinear convexity theorem [Kos73] involves the Iwasawa projection $H : G \to \mathfrak{a}$ that is defined by

$$g = ke^{H(g)}n \in KAN.$$

In representation theory, $H(g)$ often appears. Note that

$$H(g) = H(kg), \quad k \in K,$$

so

$$H(e^X k^{-1}) = H(ke^X k^{-1}) = H(e^{\operatorname{Ad}(k)X}), \quad \forall X \in \mathfrak{a}, \forall k \in K.$$

Given $X \in \mathfrak{a}$, the Kostant linear convexity theorem asserts that the map

$$F_1 : k \mapsto \text{diag}\,(\text{Ad}\,(k)X),$$

where $\text{Ad}\,(k)X = kXk^{-1}$, traces out $\text{conv}\,W \cdot X$, the convex hull of the Weyl group orbit of X. The Kostant nonlinear convexity theorem asserts that the map

$$F_2 : k \mapsto H(\exp\,(\text{Ad}\,(k)X)) = H(e^X k^{-1})$$

from K to $\mathfrak{a}$ also traces out the set $\text{conv}\,W \cdot X$. In other words,

$$F_2(K) = \text{conv}\,W \cdot X = F_1(K),$$

and π is the linearization of $\gamma := H \circ \exp$ at the origin. For example, Theorem 1.19 and Theorem 1.20 are special cases of the nonlinear convexity theorem.

Note that $H \circ \exp$ is highly nonlinear [LR91]. For example, when $G = \text{SL}_n(\mathbb{C})$,

$$(H \circ \exp)(X) = \frac{1}{2}\left(\log \Delta_1(e^{2X}), \log \frac{\Delta_2(e^{2X})}{\Delta_1(e^{2X})}, \ldots, \log \frac{\Delta_n(e^{2X})}{\Delta_{n-1}(e^{2X})}\right),$$

where $\Delta_j(e^{2X})$ is the determinant of the $j \times j$ leading principal submatrix of e^{2X}.

The following theorem of Duistermaat [Dui84] relates Kostant's linear and nonlinear convexity theorems in a nice way. The proof of Duistermaat is analytic and highly nontrivial.

Theorem 7.12. (Duistermaat) *There is a real analytic map* $\Psi : \mathfrak{p} \to K$ *such that*

(1) $\Phi_X : K \to K$ *defined by* $k \mapsto k\Psi(\text{Ad}\,k^{-1}(X))$ *is a diffeomorphism for each* $X \in \mathfrak{p}$.

(2) $\gamma(\text{Ad}\,\Psi(X)^{-1}X) = \pi(X)$ *for all* $X \in \mathfrak{p}$, *where* $\gamma = H \circ \exp$.

7.6 Thompson Theorem on Complex Symmetric Matrices

We want to draw the reader's attention to the result of Thompson in his long paper [Tho79].

Theorem 7.13. (Thompson) *Let* $d_1, \ldots, d_n \in \mathbb{C}$ *be arranged such that* $|d_1| \geqslant \cdots \geqslant |d_n|$ *and let* $s_1 \geqslant \cdots \geqslant s_n \geqslant 0$. *There is a symmetric* $A \in \mathbb{C}_{n \times n}$ *having*

$d_1, \ldots, d_n$ as its main diagonal entries and $s_1, \ldots, s_n$ as its singular values, i.e., $d \in \text{diag}\{USU^\top : U \in \text{U}(n)\}$, where $S = \text{diag}(s_1, \ldots, s_n)$, if and only if

$$\sum_{i=1}^{k} |d_i| \leqslant \sum_{i=1}^{k} s_i, \qquad \forall\, 1 \leqslant k \leqslant n,$$

$$\sum_{i=1}^{k-1} |d_i| - \sum_{i=k}^{n} |d_i| \leqslant \sum_{i=1}^{k-1} s_i + \sum_{i=k+1}^{n} s_i - s_k, \qquad \forall\, 1 \leqslant k \leqslant n,$$

$$\sum_{i=1}^{n-3} |d_i| - |d_{n-2}| - |d_{n-1}| - |d_n| \leqslant \sum_{i=1}^{n-2} s_i - s_{n-1} - s_n, \qquad \text{if } n \geqslant 3.$$

We remark that the set $\text{diag}\{USU^\top : U \in \text{U}(n)\}$ is not convex in general. However, $\text{Re diag}\{USU^\top : U \in \text{U}(n)\}$ is convex, which can be obtained by the Kostant linear convexity theorem via $\text{Sp}(n, \mathbb{R})$ [Tam99] that is the subgroup of matrices in $\text{GL}_{2n}(\mathbb{R})$ leaving invariant the exterior form

$$x_1 \wedge x_{n+1} + x_2 \wedge x_{n+2} + \cdots + x_n \wedge x_{2n}.$$

In other words, $\text{Sp}(n, \mathbb{R}) = \{A \in \text{GL}_{2n}(\mathbb{R}) : A^\top J A = J\}$, where

$$J = \begin{pmatrix} 0 & I \\ -I & 0 \end{pmatrix}.$$

It is known that

$$\mathfrak{sp}(n, \mathbb{R}) = \left\{ \begin{pmatrix} X & Y_1 \\ Y_2 & -X^\top \end{pmatrix} : Y_1^\top = Y_1,\ Y_2^\top = Y_2,\ X, Y_1, Y_2 \in \mathbb{R}_{n\times n} \right\},$$

$$K = \left\{ \begin{pmatrix} A & B \\ -B & A \end{pmatrix} : A^\top A + B^\top B = I,\ A^\top B = B^\top A,\ A, B \in \mathbb{R}_{n\times n} \right\},$$

$$\mathfrak{k} = \left\{ \begin{pmatrix} X & Y \\ -Y & X \end{pmatrix} : Y^\top = Y,\ X^\top = -X,\ X, Y \in \mathbb{R}_{n\times n} \right\},$$

$$\mathfrak{p} = \left\{ \begin{pmatrix} X & Y \\ Y & -X \end{pmatrix} : Y^\top = Y,\ X^\top = X,\ X, Y \in \mathbb{R}_{n\times n} \right\},$$

$$\mathfrak{a} = \bigoplus_{1 \leqslant j \leqslant n} \mathbb{R}(E_{jj} - E_{n+j,n+j}).$$

Through the map $\gamma : K \to \text{U}(n)$, where

$$\gamma\begin{pmatrix} A & B \\ -B & A \end{pmatrix} = A + iB,$$

we identify K with $\text{U}(n)$. The map γ preserves matrix multiplication as well as addition. In the same way, we identify $\mathfrak{k}$ with $\mathfrak{u}(n)$. We identify $\mathfrak{p}$ with S, the space of $n \times n$ complex symmetric matrices via the map $\delta : \mathfrak{p} \to S$ such that

$$\delta\begin{pmatrix} X & Y \\ Y & -X \end{pmatrix} = Y + iX, \qquad X, Y \in \mathbb{R}_{n\times n},\ X^\top = X,\ Y^\top = Y.$$

Hence $\mathfrak{a}$ is identified with the space of real diagonal matrices via δ (more precisely, the space of purely imaginary diagonal matrices). Notice that

$$\delta\left[\begin{pmatrix} A & B \\ -B & A \end{pmatrix}\begin{pmatrix} X & Y \\ Y & -X \end{pmatrix}\begin{pmatrix} A & B \\ -B & A \end{pmatrix}^\top\right] = (A+iB)(Y+iX)(A+iB)^\top.$$

Hence with these identifications, K acts on $\mathfrak{p}$, via adjoint action, such that $A \mapsto UAU^\top$, where A is complex symmetric and U is unitary. The Weyl group acts on $\mathfrak{a}$ by permutations and sign changes of the diagonal entries of matrices in $\mathfrak{a}$. The orthogonal projection $\pi : \mathfrak{p} \to \mathfrak{a}$ amounts to taking the real parts of the diagonal elements of $A \in \mathfrak{p}$. So, the fact that

$$\operatorname{Re}\operatorname{diag}\{USU^\top : U \in \mathrm{U}(n)\} = \{d \in \mathbb{C}^n : |d| \prec_w s\}$$

follows from Theorem 7.1.

Let $\mathrm{SO}(n,\mathbb{C}) = \{A \in \mathrm{GL}_n(\mathbb{C}) : A^\top A = I_n\}$ be the group of $n \times n$ complex orthogonal matrices. The subgroup $\mathrm{SO}^*(2n)$ of $\mathrm{SO}(2n,\mathbb{C})$ leaves invariant the skew Hermitian form [Hel78, p.445]

$$-z_1\bar{z}_{n+1} + z_{n+1}\bar{z}_1 - z_2\bar{z}_{n+2} + z_{n+2}\bar{z}_2 - \cdots - z_n\bar{z}_{2n} + z_{2n}\bar{z}_n.$$

In other words, $\mathrm{SO}^*(2n) = \{A \in \mathrm{SO}(2n,\mathbb{C}) : A^*JA = J\}$. It is known that

$$\mathfrak{so}^*(2n) = \left\{\begin{pmatrix} X & Y \\ -\bar{Y} & \bar{X} \end{pmatrix} : X^\top = -X,\ Y^* = Y,\ X,Y \in \mathbb{C}_{n\times n}\right\},$$

$$K = \left\{\begin{pmatrix} A & B \\ -B & A \end{pmatrix} : A^\top A + B^\top B = I,\ A^\top B = B^\top A,\ A,B \in \mathbb{R}_{n\times n}\right\},$$

$$\mathfrak{k} = \left\{\begin{pmatrix} X & Y \\ -Y & X \end{pmatrix} : X^\top = -X,\ Y^\top = Y,\ X,Y \in \mathbb{R}_{n\times n}\right\},$$

$$\mathfrak{p} = \left\{\begin{pmatrix} X & Y \\ Y & -X \end{pmatrix} : X^\top = -X,\ Y^\top = -Y,\ X,Y \in i\mathbb{R}_{n\times n}\right\},$$

$$\mathfrak{a} = i\mathbb{R}((E_{12} - E_{21}) - (E_{n+1,n+2} - E_{n+2,n+1})) \oplus$$
$$i\mathbb{R}((E_{23} - E_{32}) - (E_{n+2,n+3} - E_{n+3,n+2})) \oplus \cdots.$$

Analogously to the $\mathrm{Sp}(n,\mathbb{R})$ case, we identify K with the unitary group $\mathrm{U}(n)$ and $\mathfrak{p}$ with the space of complex skew symmetric matrices via γ and δ, respectively. Then $\mathfrak{a}$ is identified with $i \oplus_{1 \leqslant j \leqslant [n/2]} \mathbb{R}(E_{2j-1,2j} - E_{2j,2j-1})$. The group K acts on $\mathfrak{p}$, via adjoint action, in a way that $A \to UAU^\top$. The orthogonal projection amounts to taking the real part (more precisely, the imaginary part) of the vectors $d = (d_1, \ldots, d_n)$, where $d_j = (UAU^\top)_{2j-1,2j}$, $j = 1, \ldots, n$. The Weyl group acts by permutations and sign changes of the matrices in $\mathfrak{a}$.

Given a complex skew symmetric matrix A, the Autonne decomposition [HJ91] says that there exists a unitary matrix U such that

$$U^\top A U = \begin{pmatrix} 0 & s_1 \\ -s_1 & 0 \end{pmatrix} \oplus \cdots \oplus \begin{pmatrix} 0 & s_n \\ -s_n & 0 \end{pmatrix} \oplus 0,$$

where $s_1 \geqslant s_2 \geqslant s_2 \geqslant \cdots \geqslant s_n \geqslant 0$ are the singular values of A. The 1×1 block 0 vanishes if $A \in \mathbb{C}_{2n \times 2n}$. This proves the following theorem.

Theorem 7.14. *Let A be a complex skew symmetric matrix with canonical form S'. Let*

$$K(A) = \{d \in \mathbb{C}^n : d_j = (UAU^\top)_{2j-1,2j} \text{ for all } 1 \leqslant j \leqslant n \text{ and } U \in \mathrm{U}(n)\}.$$

Then $\operatorname{Re} K(A)$ and $\operatorname{Im} K(A)$ are equal to $\operatorname{conv} G(n)s$, i.e., $d \in \operatorname{Re} K(A)$ if and only if $|d| \prec_w s$.

Thompson's motivation for the study of Theorem 7.13 was a conjecture of two physicists, Tromberg and Waldenstrom, on the diagonal of symmetric unitary matrices. The diagonals are of interest to physicists because they yield probabilities in certain physical processes. The proof of Theorem 7.13 given in [Tho79] is long with 27 lemmas. It was Thompson's quest for a more conceptual approach, as he wrote, "Unfortunately, this latter theorem has a long and intricate proof, ... A proof more conceptual than that to follow would be of great interest, particularly if it should reveal the underlying geometrical properties..." and in the Note added in proof, he further wrote, "Lie theory undoubtedly affords a vehicle for establishing the above results without such an elaborate analysis of cases." We hope that Thompson's question will be answered one day.

Notes and References. This section is based on [Kos73, Tam97, Tam99]. In his 1988 Johns Hopkins Lecture Notes, Robert C. Thompson suspected that the works of Kostant [Kos73] and Eaton and Perlman [EP77] may lead to an explanation of the subtracted terms in his inequalities on singular values-diagonal entries. This turned out to be the case [Tam99].

Kostant's seminal paper was followed up by many researchers, for example, Heckman [Hec82], Atiyah [Ati82], and Guillemin and Sternberg [GS82]. Ziegler [Zie92] gave a short proof for the compact case of Kostant's linear convexity theorem by making use of representation theory and the projective embeddings of Borel-Weil-Tits.

See [HT16] for generalizations of some results in this chapter in the context of Eaton triples.

Bibliography

[AH94] T. Ando and F. Hiai. Log majorization and complementary Golden-Thompson type inequalities. *Linear Algebra Appl.*, 197/198:113–131, 1994.

[ALM04] T. Ando, C. K. Li, and R. Mathias. Geometric means. *Linear Algebra Appl.*, 385:305–334, 2004.

[And79] T. Ando. Concavity of certain maps on positive definite matrices and applications to Hadamard products. *Linear Algebra Appl.*, 26:203–241, 1979.

[APS11] J. Antezana, E. R. Pujals, and D. Stojanoff. The iterated Aluthge transforms of a matrix converge. *Adv. Math.*, 226:1591–1620, 2011.

[Ara90] H. Araki. On an inequality of Lieb and Thirring. *Lett. Math. Phys.*, 19:167–170, 1990.

[Ati82] M. F. Atiyah. Convexity and commuting Hamiltonians. *Bull. London Math. Soc.*, 14:1–15, 1982.

[Aud08] K. M. R. Audenaert. On the Araki-Lieb-Thirring inequality. *Int. J. Inf. Syst. Sci.*, 4:78–83, 2008.

[BD95] R. Bhatia and C. Davis. A Cauchy-Schwartz inequality for operators with applications. *Linear Algebra Appl.*, 223/224:119–129, 1995.

[Ber88] D. S. Bernstein. Inequalities for the trace of matrix exponentials. *SIAM J. Matrix Anal. Appl.*, 9:156–158, 1988.

[Ber09] D. S. Bernstein. *Matrix Mathematics: Theory, Facts, and Formulas (2nd ed.)*. Princeton University Press, 2009.

[BH06] R. Bhatia and J. Holbrook. Riemannian geometry and matrix geometric means. *Linear Algebra Appl.*, 413:594–618, 2006.

[Bha97] R. Bhatia. *Matrix Analysis*. Springer-Verlag, 1997.

[Bha07] R. Bhatia. *Positive Definite Matrices*. Princeton University Press, 2007.

[Bha13]   R. Bhatia. The Riemannian mean of positive matrices. In *Matrix Information Geometry*, pages 35–51. Springer, 2013.

[Bou68]   N. Bourbaki. *Lie Groups and Lie Algebras, Chapter 4–6*. Springer, 1968.

[BtD85]   T. Bröcker and T. tom Dieck. *Representations of Compact Lie Groups*. Springer-Verlag, 1985.

[CFKK82] J. E. Cohen, S. Friedland, T. Kato, and F. P. Kelly. Eigenvalue inequalities for products of matrix exponentials. *Linear Algebra Appl.*, 45:55–95, 1982.

[CL83]   N. N. Chan and K. H. Li. Diagonal elements and eigenvalues of a real symmetric matrix. *J. Math. Anal. Appl.*, 91:562–566, 1983.

[Coh88]   J. E. Cohen. Spectral inequalities for matrix exponentials. *Linear Algebra Appl.*, 111:25–28, 1988.

[Cor87]   H. Cordes. *Spectral Theory of Linear Differential Operators and Comparison Algebras*. Cambridge University Press, 1987.

[Dui84]   J. J. Duistermaat. On the similarity between the Iwasawa projection and the diagonal part. *Soc. Math. France (N. S.)*, 15:129–138, 1984.

[EP77]   M. Eaton and M. Perlman. Reflection groups, generalized Schur functions, and the geometry and majorization. *Ann. Probab.*, 5:829–860, 1977.

[Fan49]   K. Fan. On a theorem of Weyl concerning eigenvalues of linear transformations, I. *Proc. Nat. Acad. Sci. USA*, 35:652–655, 1949.

[Fan51]   K. Fan. Maximum properties and inequalities for the eigenvalues of completely continuous operators. *Proc. Nat. Acad. Sci. USA*, 37:760–766, 1951.

[FH55]   K. Fan and A. J. Hoffman. Some metric inequalities in the space of matrices. *Proc. Amer. Math. Soc.*, 6:111–116, 1955.

[FH90]   T. Furuta and J. Hakeda. Applications of norm inequalities equivalent to Löwner-Heinz theorem. *Nihonkai Math. J.*, 1:11–17, 1990.

[Fie74]   M. Fiedler. Additive compound matrices and an inequality for eigenvalues of symmetric stochastic matrices. *Czechoslovak Math. J.*, 24:392–402, 1974.

[FT14]   P. J. Forrester and C. J. Thompson. The Golden-Thompson inequality: Historical aspects and random matrix applications. *J. Math. Phys.*, 55:023503, 2014.

Bibliography

[Gol65] S. Golden. Lower bounds for the Helmholtz function. *Phys. Rev.*, 137:1127–1128, 1965.

[GS82] V. Guillemin and S. Sternberg. Convexity properties of the moment mapping. *Invent. Math.*, 67:491–513, 1982.

[Hal82] P. R. Halmos. *A Hilbert Space Problem Books (2nd ed.)*. Springer-Verlag, 1982.

[Hal03] B. C. Hall. *Lie Groups, Lie Algebras, and Representations: An Elementary Introduction*. Springer, 2003.

[Hec82] G. J. Heckman. Projections of orbits and asymptotic behavior of multiplicities for compact connected Lie groups. *Invent. Math.*, 67:333–356, 1982.

[Hel78] S. Helgason. *Differential Geometry, Lie Groups, and Symmetric Spaces*. Academic Press, 1978.

[Hia97] F. Hiai. Log-majorizations and norm inequalities for exponential operators. In *Linear Operators*, volume 38 of *Banach Center Publications*, pages 119–181. Polish Acad. Sci., 1997.

[HJ91] R. A. Horn and C. R. Johnson. *Topics in Matrix Analysis*. Cambridge University Press, 1991.

[HJ13] R. A. Horn and C. R. Johnson. *Matrix Analysis (2nd ed.)*. Cambridge University Press, 2013.

[HK10] H. Huang and S. Kim. On Kostant's partial order on hyperbolic elements. *Linear and Multilinear Algebra*, 58:783–788, 2010.

[Hor54a] A. Horn. Doubly stochastic matrices and the diagonal of a rotation of matrix. *Amer. J. Math.*, 76:620–630, 1954.

[Hor54b] A. Horn. On the eigenvalues of a matrix with prescribed singular values. *Proc. Amer. Math. Soc.*, 5:4–7, 1954.

[How83] R. Howe. Very basic Lie theory. *Amer. Math. Monthly*, 90:600–623, 1983.

[HP93] F. Hiai and D. Petz. The Golden-Thompson trace inequality is complemented. *Linear Algebra Appl.*, 181:153–185, 1993.

[HT10] H. Huang and T. Y. Tam. On the Gelfand-Naimark decomposition of a nonsingular matrix. *Linear and Multilinear Algebra*, 58:27–43, 2010.

[HT16] W. C. Hill and T. Y. Tam. Derivatives of orbital function and an extension of Berezin-Gel'fand's theorem. *Spec. Matrices*, 4:333–349, 2016.

[Hum72]   J. E. Humphreys. *Introduction to Lie Algebras and Representation Theory.* Springer-Verlag, 1972.

[JN90]   C. R. Johnson and P. Nylen. Yamamoto's theorem for generalized singular values. *Linear Algebra Appl.*, 128:147–158, 1990.

[JN93]   C. R. Johnson and P. Nylen. Erratum: Yamamoto's theorem for generalized singular values, [Linear Algebra Appl., 128:147–158 (1990)]. *Linear Algebra Appl.*, 180:4, 1993.

[Kat61]   T. Kato. A generalization of the Heinz inequality. *Proc. Japan Acad.*, 37:305–308, 1961.

[Kit93]   F. Kittaneh. Norm inequalities for fractional powers of positive operators. *Lett. Math. Phys.*, 27:279–285, 1993.

[Kna02]   A. W. Knapp. *Lie Groups Beyond An Introduction (2nd ed.).* Birkhäuser, 2002.

[Kos73]   B. Kostant. On convexity, the Weyl group and the Iwasawa decomposition. *Ann. Sci. École Norm. Sup. (4)*, 6:413–455, 1973.

[KT99]   A. Knutson and T. Tao. The honeycomb model of $GL_n(\mathbb{C})$ tensor products, I, Proof of the saturation conjecture. *J. Amer. Math. Soc.*, 12:1055–1090, 1999.

[KT01]   A. Knutson and T. Tao. Honeycombs and sums of Hermitian matrices. *Notices Amer. Math. Soc.*, 48:175–186, 2001.

[Lan99]   S. Lang. *Fundamentals of Differential Geometry.* Springer, 1999.

[Lee13]   J. Lee. *Introduction to Smooth Manifolds (2nd ed.).* Springer-Verlag, 2013.

[Len71]   A. Lenard. Generalization of the Golden-Thompson inequality $\operatorname{tr}(e^A e^B) \geq \operatorname{tr} e^{A+B}$. *Indiana Univ. Math. J.*, 21:457–467, 1971.

[Lia04]   M. Liao. *Lévy Processes in Lie Groups.* Cambridge University Press, 2004.

[Lim12]   Y. Lim. Factorizations and geometric means of positive definite matrices. *Linear Algebra Appl.*, 437:2159–2172, 2012.

[Liu17]   X. Liu. Generalization of some inequalities for matrix exponentials to Lie groups. *J. Lie Theory*, 27:185–192, 2017.

[LL01]   J. Lawson and Y. Lim. The geometric mean, matrices, metrics, and more. *Amer. Math. Monthly*, 108:797–812, 2001.

[LLT14]   M. Liao, X. Liu, and T. Y. Tam. A geometric mean for symmetric spaces of noncompact type. *J. Lie Theory*, 24:725–736, 2014.

Bibliography

[LR91] J-H. Lu and T. Ratiu. On the nonlinear convexity theorem of Kostant. *J. Amer. Math. Soc.*, 4:349–363, 1991.

[LT76] E. Lieb and W. Thirring. Inequalities for the moments of the eigenvalues of the Schrödinger Hamiltonian and their relation to Sobolev inequalities. In *Studies in Mathematical Physics*, pages 269–303. Princeton University Press, 1976.

[LT14] X. Liu and T. Y. Tam. Extensions of three matrix inequalities to semisimple Lie groups. *Spec. Matrices*, 2:148–154, 2014.

[Mar73] M. Marcus. *Finite Dimensional Multilinear Algebra I*. Marcel Dekker, 1973.

[Mar75] M. Marcus. *Finite Dimensional Multilinear Algebra II*. Marcel Dekker, 1975.

[McI79] A. McIntosh. Heinz inequalities and perturbation of spectral families. *Macquarie Mathematics Reports 79-0006, Macquarie University*, 1979.

[Mer97] R. Merris. *Multilinear Algebra*. Gordon and Breach Science Publishers, 1997.

[Mir60] L. Mirsky. Symmetric gauge functions and unitarily invariant norms. *Quart. J. Math. Oxford*, 11:50–59, 1960.

[Moa05] M. Moakher. A differential geometric approach to the geometric mean of symmetric positive-definite matrices. *SIAM J. Matrix Anal. Appl.*, 26:735–747, 2005.

[MOA11] A. W. Marshall, I. Olkin, and B. C. Arnold. *Inequalities: Theory of Majorization and its Applications (2nd ed.)*. Springer, 2011.

[NR90] P. Nylen and L. Rodman. Approximation numbers and Yamamoto's theorem in Banach algebras. *Integral Equations Operator Theory*, 13:728–749, 1990.

[OV94] A. L. Onishchik and E. B. Vinberg. *Lie Groups and Lie Algebras III: Structure of Lie Groups and Lie Algebras*. Springer-Verlag, 1994.

[Pet94] D. Petz. A survey of certain trace inequalities. In *Functional Analysis and Operator Theory*, volume 30 of *Banach Center Publications*, pages 287–298. Polish Acad. Sci., 1994.

[Roi99] M. Roitman. A short proof of the Jordan decomposition theorem. *Linear and Multilinear Algebra*, 46:245–247, 1999.

Bibliography

[Sch23]   I. Schur. Über eine Klasse von Mittelbildungen mit Anwendungen auf der Determinantentheorie. *Sitzungsberichte der Berlinear Mathematischen Gesellschaft*, 22:9–20, 1923.

[Sch70]   B. Schwarz. Totally positive differential systems. *Pacific J. Math.*, 32:203–229, 1970.

[Ser10]   D. Serre. *Matrices: Theory and Applications (2nd ed.)*, volume 216 of *Graduate Texts in Mathematics*. Springer, 2010.

[Sim79]   B. Simon. *Trace Ideals and Their Applications*, volume 35 of *London Mathematical Society Lecture Note Series*. Cambridge University Press, 1979.

[Sin76]   F. Y. Sing. Some results on matrices with prescribed diagonal elements and singular values. *Canad. Math. Bull.*, 19:89–92, 1976.

[So92]   W. So. Equality cases in matrix exponential inequalities. *SIAM J. Math. Anal. Appl.*, 13:1154–1158, 1992.

[Sym65]   K. Symanzik. Proof of refinements of an inequality of Feynmann. *J. Math. Phys.*, 6:1155–1156, 1965.

[Tam97]   T. Y. Tam. Kostant's convexity theorem and the compact classical groups. *Linear and Multilinear Algebra*, 43:87–113, 1997.

[Tam99]   T. Y. Tam. A Lie-theoretic approach to Thompson's theorems on singular values-diagonal elements and some related results. *J. London Math. Soc.*, 60:431–448, 1999.

[Tam08]   T. Y. Tam. Heinz-Kato's inequalities for Lie groups. *J. Lie Theory*, 18:919–931, 2008.

[Tam10a]   T. Y. Tam. A. Horn's result on matrices with prescribed singular values and eigenvalues. *Electron. J. Linear Algebra*, 21:25–27, 2010.

[Tam10b]   T. Y. Tam. Some exponential inequalities for semisimple Lie groups. In *Operators, Matrices and Analytic Functions*, volume 202 of *Operator Theory: Advances and Applications*, pages 539–552. Birkhäuser Verlag, 2010.

[Tho65]   C. J. Thompson. Inequality with applications in statistical mechanics. *J. Mathematical Phys.*, 6:1812–1813, 1965.

[Tho71]   C. J. Thompson. Inequalities and partial orders on matrix spaces. *Indiana Univ. Math. J.*, 21:469–480, 1971.

[Tho77]   R. C. Thompson. Singular values, diagonal elements, and convexity. *SIAM J. Appl. Math.*, 32:39–63, 1977.

[Tho79]   R. C. Thompson. Singular values and diagonal elements of complex symmetric matrices. *Linear Algebra Appl.*, 26:65–106, 1979.

[Tho92]   R. C. Thompson. High, low, and quantitative roads in linear algebra. *Linear Algebra Appl.*, 162–164:23–64, 1992.

[vN37]   J. von Neumann. Some matrix-inequalities and metrization of metric-space. *Tomsk. Univ. Rev.*, 1:286–300, 1937.

[War83]   F. W. Warner. *Foundations of Differentiable Manifolds and Lie Groups*. Springer-Verlag, 1983.

[Wey49]   H. Weyl. Inequalities between the two kinds of eigenvalues of a linear transformation. *Proc. Nat. Acad. Sci. U. S. A.*, 35:408–411, 1949.

[Yam67]   T. Yamamoto. On the extreme values of the roots of matrices. *J. Math. Soc. Japan*, 19:173–178, 1967.

[Yam12]   T. Yamazaki. The Riemannian mean and matrix inequalities related to the Ando-Hiai inequality and chaotic order. *Oper. Matrices*, 6:577–588, 2012.

[Zha02]   X. Zhan. *Matrix Inequality*, volume 1790 of *Lecture Notes in Mathematics*. Springer-Verlag, 2002.

[Zha11]   F. Zhang. *Matrix Theory: Basic Results and Techniques (2nd ed.)*. Universitext. Springer, 2011.

[Zie92]   F. Ziegler. On the Kostant convexity theorem. *Proc. Amer. Math. Soc.*, 115:1111–1113, 1992.

Index

adjoint group, 47
adjoint representation, 44, 47

Cartan involution
    Lie algebra, 51
    Lie group, 53
Cartan subalgebra, 48
closed subgroup, 46
complex structure, 49
complexification
    Lie algebra, 50
    vector space, 49
compound matrix
    additive, 27
    multiplicative, 26
conjugation, 50
convex hull, 15
coordinate chart, 42

derivation, 43
diffeomorphism, 42
differential, 43

elliptic, 62
exponential map, 45

Fan dominance theorem, 22

geometric mean, 109

homomorphism, 44
    derived, 45
    smooth, 45
hyperbolic, 62

ideal, 44
immersion, 43
inequality

Araki-Lieb-Thirring, 75
Bernstein, 76
Cordes, 71
Golden-Thompson, 67

Jordan basis, 7

Killing form, 47

Lie algebra, 44
    abelian, 47
    compact, 47
    general linear algebra, 44
    nilpotent, 47
    reductive, 47
    semisimple, 47
    simple, 47
    solvable, 47
Lie algebra decomposition
    Cartan, 5, 51
    Iwasawa, 6, 58
    root space, 48, 55
Lie bracket, 44
Lie group, 45
    closed linear group, 45
    general linear group, 45
Lie group decomposition
    KA_+K, 4, 61
    Bruhat, 13
    Cartan, 4, 53
    complete multiplicative Jordan decomposition, 62
    Iwasawa, 5, 58
Lie product formula, 24, 82
Löwner order, 18

majorization
    log, 15

weak, 14
weakly log, 14
manifold
    smooth, 42
    topological, 41
matrix
    doubly stochastic, 15
    elliptic, 6
    hyperbolic, 6
    nilpotent, 6
    real semisimple, 6
    semisimple, 6
    unipotent, 6
matrix decomposition
    $L\omega U$, 10
    LU, 9
    QR, 4
    Cartesian, 5
    Cholesky, 5
    Gelfand-Naimark, 10
    Jordan (additive), 6
    Jordan (complete multiplicative), 8
    Jordan (multiplicative), 8
    polar, 4
    singular value, 4
matrix exponential map, 23
matrix group
    general linear group, 3
    orthogonal group, 3
    special linear group, 3
    unitary group, 3

nilpotent, 48, 61
norm, 19
    Euclidean, 19
    Frobenius, 19
    Ky Fan k, 20
    matrix, 19
    Schatten p, 20
    spectral, 19
    symmetric gauge invariant, 20
    unitarily invariant, 19
normal, 54

one-parameter subgroup, 45

partial order, 13
preorder, 13

real form, 50
real semisimple, 61
realification
    Lie algebra, 50
    vector space, 49
realizable, 34
root, 48, 55
    positive, 56
root system, 48

semisimple, 48
singular value, 4
    real, 87
smooth chart, 42
smooth function, 42
smooth map, 42
smooth structure, 42
subalgebra, 44
subgroup, 46
submanifold, 43
symmetric gauge function, 20

tangent bundle, 43
tangent space, 43
tangent vectors, 43
toral subalgebras, 48

unipotent, 62

vector field, 43

Weyl chamber, 56
    fundamental, 56
Weyl group, 59, 60